Deepen Your Mind

Deepen Your Mind

序

從學生時期開始，我對於「聊天機器人」就有一種莫名的熱情，早在十幾年前 MSN 盛行的年代，我就已透過古老的 Flash 技術，設計出了一個模擬 MSN 介面的聊天機器人，不僅能進行基本的日常對話，還能傳送當時最夯的震動、表情圖片和大頭貼，這個機器人甚至幫我完成了研究所的畢業論文（碩博士論文搜尋：智慧型代理人之表情設計研究）。

隨著 MSN 和 Flash 消失，當初的機器人從此長眠，但也因此在我心中種下了「AI 聊天機器人」的種子，直到 LINE 的出現，又引燃了我的這份熱情，由於 LINE 提供的開發文件相當完整（雖然有些仍然得靠自己東拼西湊），讓開發者可以很方便地進行開發，所以越來越多的公司企業，都會讓自己擁有一個客服或宣傳的 LINE BOT，許多政治人物、藝人媒體或團購，也都會利用 LINE BOT 作為自己行銷的管道。

最開始接觸 LINE BOT 時，我使用 Google Apps Script 進行開發，在一兩年裡實作出好幾個 LINE BOT，有些變成了公司的產品，有些串接物聯網裝置成為了智慧家庭管家，有些甚至成為了我 LINE BOT 授課時的課堂經典範例。接觸了 Python 之後，發現透過 Python 更容易實現爬蟲、影像處理的功能，所以開始利用 Python 開發 LINE BOT，這也成為了出版這本書的契機。

這本書完整紀錄了使用 Python 開發 LINE BOT 的過程，從註冊官方帳號、註冊開發者服務開始，一直介紹到 LINE Message API 的相關用法、LINE 訊息的解析以及開發 LINE BOT 的眉眉角角，最後還會使用 Google Cloud Functions 打造 24 小時不間斷的機器人服務，使用 Google Dialogflow 打造能理解自然語意的 AI 聊天機器人，整本書盡可能減少累贅的文字敘述，將重點擺在實際程式碼的操作過程，不論是老手新手，只要跟著書中的步驟，就能輕鬆掌握 LINE BOT 的開發精髓，打造自己獨一無二的 AI 聊天機器人。

目錄

Part 4 解析 LINE 訊息

Part 5 傳送 LINE 訊息的方法

Part 6 傳送不同類型的 LINE 訊息

Part 7 實作 LINE 氣象機器人

Part 8 串接 Dialogflow 打造聊天機器人

Part 9 使用 LINE Notify 推播通知

Part 10 使用 Google Cloud Functions

附錄 其他參考資訊

1

認識 LINE BOT

前言

LINE 在全世界擁有上千萬的用戶,在台灣幾乎每個使用手機的人,都會註冊一個 LINE 的帳號,不論公家單位還是私人企業,也都會使用 LINE 的官方帳號來與使用者保持聯繫。隨著資訊科技的發展,近年來各個企業也逐漸打造自己的 LINE BOT,除了被動的處理客服諮詢,還能透過 LINE BOT 主動推播各種行銷活動資訊,透過與使用者的即時互動,打造優質的品牌形象。

1-1　什麼是 LINE BOT

LINE BOT 表示 LINE 機器人，如果搭配聊天功能，亦可稱為 LINE 聊天機器人，大多數 LINE 的官方帳號的後台都會串接一個 LINE BOT 服務，透過 LINE 多元的功能（圖文選單、樣板選單、輪播訊息 ... 等），能夠提供給使用者非常方便的功能體驗。

一個最簡單的 LINE BOT，可以從註冊官方帳號開始，註冊了 LINE 的官方帳號後，從管理介面中就可以進行最簡單的 LINE BOT 聊天操作，例如設定群發訊息、加入好友時自動通知訊息、定時發送訊息 ... 等。

LINE 官方帳號：https://tw.linebiz.com/login/

除了 LINE 本身提供的許多官方帳號，例如 LINE 購物、LINE Today、LINE 運動 …… 等，各個公司企業或政府機關，也都會使用 LINE BOT 作為官方宣傳的媒介，或是客服溝通的管道，這些不外乎都是 LINE BOT 的基本應用。

1-2　LINE BOT 在生活中的應用

除了官方帳號會串接 LINE BOT，如果要讓 LINE BOT 擁有更多的功能，就需要撰寫程式去串接 LINE Developer 所提供的操作方法，透過各種方法的互相搭配，就能打造自己專屬的 LINE BOT，例如能夠理解自然語意的聊天機器人、記帳機器人、購票機器人、天氣通知機器人 … 等。

近期比較有名的應用，不外乎是一些搭配網路爬蟲的機器人，例如全台快篩低價販賣所機器人可以幫助大家查詢哪裡可以買快篩，敗口罩機器人可以幫助大家查詢哪裡可以買口罩。

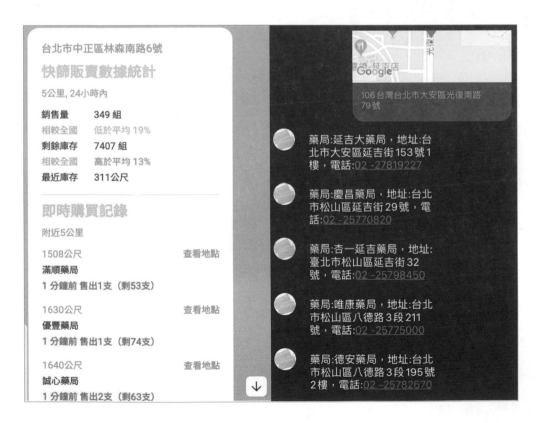

小結

在目前的時代裡，LINE 已經和生活息息相關，不論是政府機關、公司企業、廣告宣傳，甚至是團購代購，都想要與 LINE 的功能沾上邊，既然脫離不了 LINE 的魔掌？！不如就學點 Python 程式，透過 Python 打造自己的 LINE BOT，徹底發揮 LINE 百分之兩百的功能吧！

Note

2

建立 LINE BOT

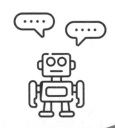

前言

如果要開發一個 LINE BOT（LINE 聊天機器人），必須要先成為 LINE 的開發者，成為開發者後就能開始使用 LINE Developer 的 Message API 功能進行訊息的通訊，這個章節將會介紹如何成為 LINE 的開發者，並建立一個 LINE BOT 的 Channel。

2-1 註冊並登入 LINE Developer

前往 LINE Developers 網站，使用個人帳號登入。

LINE Developers：https://developers.line.biz/zh-hant/

2-2 建立 Provider

登入後，點選 Create 建立一個 Provider (表示開發 LINE BOT 後所在的群組)

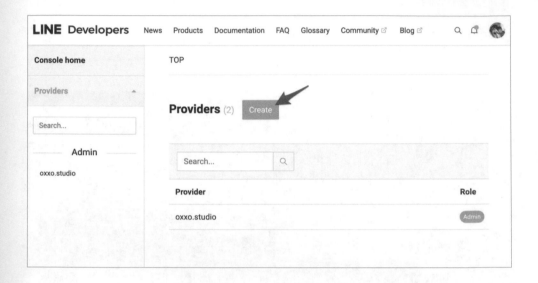

輸入名稱，點擊 Create 就能建立 Provider。

Create a new provider

Provider name ⑦   test

✓ Don't leave this empty
✓ Don't use special characters (4-byte Unicode)
✓ Enter no more than 100 characters

A provider is an individual developer, company, or organization that provides services. For more details, see the documentation ⌐ .

Cancel   Create

2-3　建立 Channel

建立 Provider 後，選擇 Channels 頁籤，點選「Create a new channel」建立頻道，一個頻道表示一個聊天機器人。

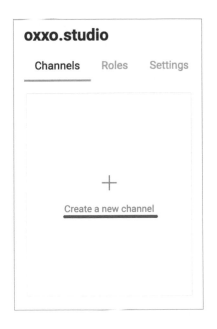

選擇「Message API」，建立聊天訊息專屬的頻道，建立時需要輸入頻道的名稱和描述述，以及使用下拉選擇所在的地區與類別。

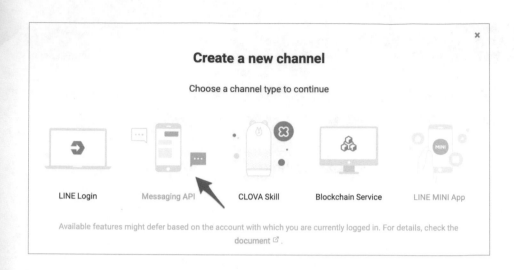

建立完成後，在 Provider 裡就會看到出現了剛剛建立的 Channel。

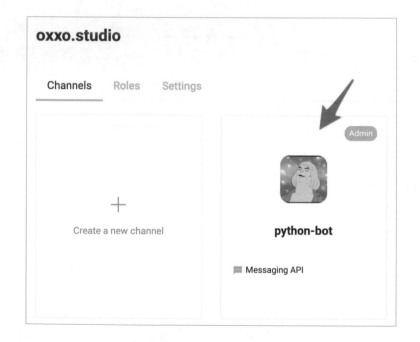

2-4　LINE 官方帳號設定

建立 Channel 後，前往「LINE 官方帳號管理頁面」。

LINE 官方帳號管理頁面：https://tw.linebiz.com/login/

點擊「登入管理頁面」。

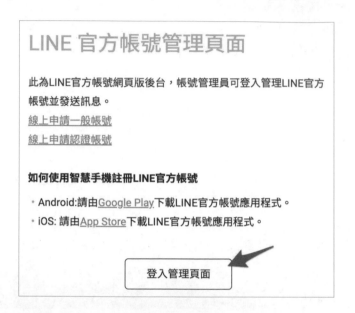

登入後，在帳號一覽的清單裡，就能看見剛剛建立的 Channel 自動變成了一個官方帳號。

點擊進入官方帳號，上方可以看到「回應模式：聊天機器人」，表示會透過自己撰寫的程式進行串訊息的功能。

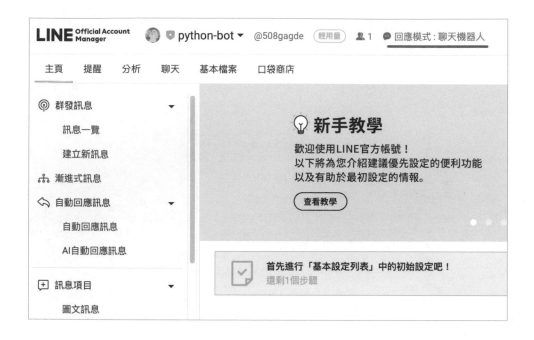

選擇「自動回應訊息」，「關閉」自動回應訊息的狀態，避免每次跟 LINE BOT 聊天時，都會跳出自動回應的訊息。

2-5 加入 LINE BOT 為好友

完成後，從「設定」畫面中找到 LINE BOT 的 ID，就能使用 LINE 增加好友的功能，將 LINE BOT 增加為好友。

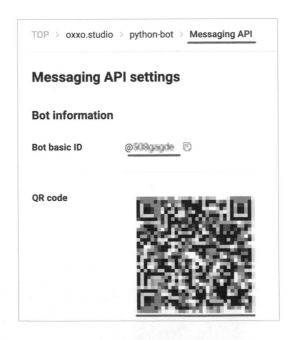

除了從官方帳號管理頁面增加 LINE BOT 為好友，也可以從 LINE Developer 控制台的 Message API 頁籤，找到 LINE BOT 的 ID 和 QRCode，使用 LINE 掃描 QRCode 增加好友的功能，掃描 QRCode 就能將 LINE BOT 加為好友。

加入後，LINE 裡面就會看到出現第一次加入好友的歡迎訊息。

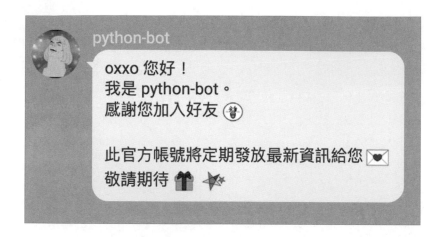

 小結

在 這 個 章 節 裡，已 經 使 用 LINE 開 發 者 的 身 份 建 立 了 一 個 LINE BOT Channel，同時也會自動產生一個「LINE 官方帳號」，有了這個 LINE BOT Channel 和官方帳號之後，就能開始正式進行聊天機器人程式的開發。

Note

3

開發環境設定 &
串接 LINE BOT

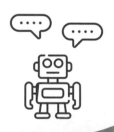

前言

建立完成 LINE BOT Channel 和 LINE 官方帳號之後，就要開始準備 LINE
BOT 的開發環境，接下來的幾個章節會引導大家使用本機環境、Google
Colab 以及 Cloud Functions，一次掌握三個 LINE BOT 開發環境的建構技巧
(可按照個人喜好，選擇適合自己的開發環境)，最後會用基本的範例程式串
接 LINE BOT。

本章節的範例程式碼：

https://github.com/oxxostudio/book-code/tree/master/linebot/ch3

3-1 認識 Webhook

Python 的開發環境最重要的就是要建立 Webhook，讓 LINE 聊天的訊息能夠透過 Webhook 傳遞到用 Python 編輯的程式去做處理。Webhook 指的是一個「**網址**」，透過伺服器建立 Webhook 網址後，**串接 Webhook 的位置就能使用 HTTP 的 POST 方法，向伺服器傳送或接收特定的資料。**

當使用者在 LINE 聊天室裡跟 LINE BOT 聊天，會發生下列的步驟：

▶Step 1　向使用 Message API 所建立的 LINE BOT 發送訊息。

▶Step 2　**訊息透過 Webhook 傳遞到使用者部署 Python 程式的伺服器。**

▶Step 3　根據 Python 程式的邏輯，處理訊息。

▶Step 4　**透過 Webhook 回傳結果到 LINE BOT。**

▶Step 5　LINE BOT 發送訊息到 LINE 聊天室裡。

下圖為 LINE Developer 官方網站的 Message API 原理圖

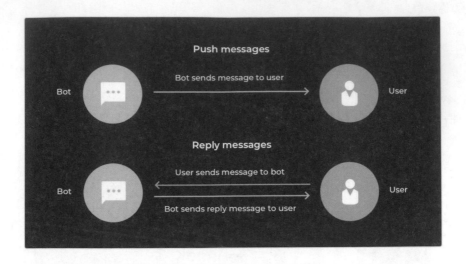

下圖為發送訊息給 LINE BOT 時的流程圖

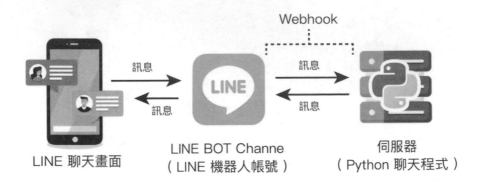

3-2　取得 LINE Channel access token 和 Channel secret

在 LINE Developer 控制台裡，進入自己建立的 LINE Channel，前往 Basic settings 頁籤，能夠看到 Channel secret (如果沒有看到或需要重新產生，點擊 Issue 就能產生)。

前往 Messaging API 頁籤，就能找到 Channel access token，它和 Channel secret 都是串接 LINE Channel 必須的金鑰 (如果沒有看到或需要重新產生，點擊 Reissue 就能產生)。

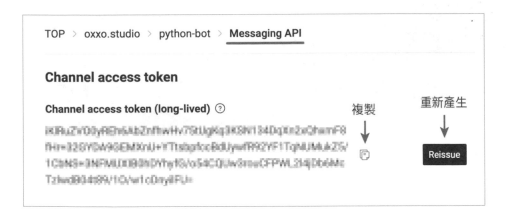

3-3 註冊 ngrok 服務

在開發階段時，常常是使用本機的伺服器，無法真正在外界進行測試，然而透過免費的 ngrok 服務，能夠將本機環境對應到一個 ngrok 網址，公開在整個網際網路中，由於是公開網址，就能真正在外界進行測試。

前往 ngrok 的網站，註冊帳號並登入。

ngrok 網站：https://ngrok.com/

登入後，從左側選單點擊 Your Authtoken，會出現一段串接 ngrok 服務所使用的 token (點擊網頁最下方 reset token 按鈕可以重設 token)。

Your Authtoken

This is your personal Authtoken. Use this to authenticate the ngrok agent that you downloaded.

```
25PfE4iyWafGGLtdbMjz9dsbP0Z_4iyWafGGLtdbMjz9dsv<
```
 Copy

3-4　建立 Webhook（本機環境）

如果是使用本機開發環境（Anaconda Jupyter 或 Python 虛擬環境），在註冊 ngrok 完成並取得 ngrok Authtoken 後，前往 ngrok 的下載頁面，根據自己電腦作業系統，使用終端機的命令下載安裝，或下載對應的安裝檔進行安裝。

下載 ngrok：https://ngrok.com/download

安裝後，開啟終端機，使用命令輸入註冊 ngrok 後取得的 token（將 <token>
換成 ngrok 的 Authtoken）。

```
ngrok authtoken <token>
```

輸入 token 後，繼續使用命令，將本機環境的埠號 port 對應到 ngrok 公開網
址（將 <token> 換成自己的 port，如果是使用 Flask 建構的服務，port 預設
為 5000）。

```
ngrok http <port>
```

完成後，就會看到終端機裡出現 ngrok 的公開網址（特別注意，每次重新輸
入執行後，網址都會改變！）

```
ngrok by @inconshreveable                               (Ctrl+C to quit)

Session Status          online
Account                 ohha12345 (Plan: Free)
Update                  update available (version 2.3.40, Ctrl-U to update
Version                 2.3.35
Region                  United States (us)
Web Interface           http://127.0.0.1:4040
Forwarding              http://9dcb-114-40-121-52.ngrok.io -> http://local
Forwarding              https://9dcb-114-40-121-52.ngrok.io -> http://loca

Connections             ttl     opn     rt1     rt5     p50     p90
                        3       0       0.00    0.00    0.01    0.01

HTTP Requests
-------------

GET /                   200 OK
GET /favicon.ico        404 NOT FOUND
GET /                   200 OK
```

使用本機的 Python 編輯器，輸入 pip install Flask 安裝 Flask 函式庫，如果是 Anaconda Jupyter 已經預設安裝，接著執行下方的程式碼，會開啟一個本機網頁服務，網址為 127.0.0.1:5000（Flask 函式庫教學參考：https://steam. oxxostudio.tw/category/python/example/flask.html）。

```python
from flask import Flask

app = Flask(__name__)

@app.route("/<name>")
def home(name):
    return f"<h1>hello {name}</h1>"

app.run()
```
（範例程式碼：ch3/code01.py）

打開瀏覽器，輸入 127.0.0.1:5000/oxxo，畫面中就會出現 hello oxxo 的文字，但這個網址只有本機瀏覽器能夠使用，外部無法使用。

由於 5000 的埠號已經和 ngrok 串接，所以**輸入剛剛透過 ngrok 產生的公開網址**，就會看到一模一樣的結果，而這個網址，不論在任何地方，都能正常讀取，進行到這個步驟，表示已經可以順利在本機開發環境，建立預備連接 LINE BOT 的 Webhook。

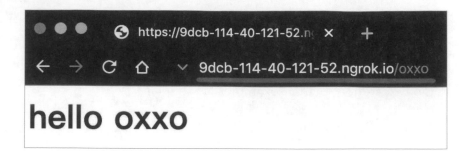

使用 ctrl + c（或 command + c）停止伺服器的程式，在終端機輸入指令安裝 line-bot-sdk 函式庫（如果使用 Anaconda Jupyter，pip 前方需要加上！變成 !pip）。

```
pip install line-bot-sdk
```

修改程式，**填入將剛剛註冊 LINE BOT 所取得的 LINE Channel access token 和 LINE Channel secret**（後續章節會介紹原理，在此處先依樣畫葫蘆撰寫，確認和 LINE BOT 可以正常串接）

```python
from flask import Flask, request

# 載入 json 標準函式庫，處理回傳的資料格式
import json

# 載入 LINE Message API 相關函式庫
from linebot import LineBotApi, WebhookHandler
from linebot.exceptions import InvalidSignatureError
from linebot.models import MessageEvent, TextMessage, TextSendMessage

app = Flask(__name__)

@app.route("/", methods=['POST'])
def linebot():
    # 取得收到的訊息內容
    body = request.get_data(as_text=True)
    try:
```

```python
        # json 格式化訊息內容
        json_data = json.loads(body)
        access_token = '你的 LINE Channel access token'
        secret = '你的 LINE Channel secret'
        # 確認 token 是否正確
        line_bot_api = LineBotApi(access_token)
        # 確認 secret 是否正確
        handler = WebhookHandler(secret)
        # 加入回傳的 headers
        signature = request.headers['X-Line-Signature']
        # 綁定訊息回傳的相關資訊
        handler.handle(body, signature)
        # 取得回傳訊息的 Token
        tk = json_data['events'][0]['replyToken']
        # 取得 LINe 收到的訊息類型
        type = json_data['events'][0]['message']['type']
        if type=='text':
            # 取得 LINE 收到的文字訊息
            msg = json_data['events'][0]['message']['text']
            # 印出內容
            print(msg)
            reply = msg
        else:
            reply = '你傳的不是文字呦～'
        print(reply)
        # 回傳訊息
        line_bot_api.reply_message(tk,TextSendMessage(reply))
    except:
        # 如果發生錯誤，印出收到的內容
        print(body)
    # 驗證 Webhook 使用，不能省略
    return 'OK'

if __name__ == "__main__":
    app.run()
```

（ 範例程式碼：ch3/code02.py ）

完成後執行程式，就會啟動本機環境的伺服器，在 ngrok 保持啟動的狀態下，複製 ngrok 所產生的公開網址，前往「**LINE Developer 控制台**」，進入自己建立的 LINE BOT Channel，**在 Message API 頁籤裡找到 Webhook setting 選項，貼上 ngrok 網址**（注意要使用 https，圖片網址為示意圖，請使用自己 ngrok 產生的網址）。

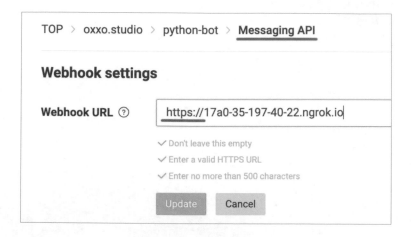

完成後，點擊「Verify」進行驗證，如果**出現 Success 文字表示已經可以正常串接**。

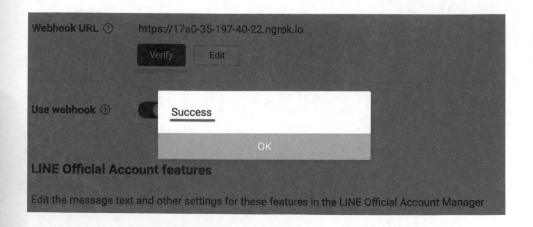

開啟 LINE，找到 LINE BOT 的帳號，傳送訊息給 LINE BOT，LINE BOT 就會
回應相同的訊息。

3-5 建立 Webhook (Google Colab)

除了本機的開發環境，也可以使用 Google Colab 作為線上開發與測試的工具，不過受限於 Colab 是無主機的開發環境，只要閒置過久或重新啟動 Colab，都需要重新安裝 ngrok (如果習慣本機環境開發，可不用閱讀此章節)。

開啟一個 Colab 的新專案，輸入並執行下方程式碼，就會連動自己帳號的 Google Drive，並在指定目錄下建立 ngrok 相關目錄 (過程中會出現彈出視窗，要求允許權限，選擇對應的帳號允許權限即可)。

```
from google.colab import drive
drive.mount('/content/drive', force_remount=True)

!mkdir -p /drive
!mount --bind /content/drive/My\ Drive /drive
!mkdir -p /drive/ngrok-ssh
!mkdir -p ~/.ssh
```
(範例指令：ch3/cmd1.txt)

```
from google.colab import drive
drive.mount('/content/drive', force_remount=True)

!mkdir -p /drive
#umount /drive
!mount --bind /content/drive/My\ Drive /drive
!mkdir -p /drive/ngrok-ssh
!mkdir -p ~/.ssh

Mounted at /content/drive
```

輸入並執行下方程式碼，會下載 ngrok 的 ssh 檔案，並自動解壓縮放到指定的目錄下，目的在於要讓 ngrok 網站認證身份。

```
!mkdir -p /drive/ngrok-ssh
%cd /drive/ngrok-ssh
!wget https://bin.equinox.io/c/4VmDzA7iaHb/ngrok-stable-linux-amd64.
zip -O ngrok-stable-linux-amd64.zip
!unzip -u ngrok-stable-linux-amd64.zip
!cp /drive/ngrok-ssh/ngrok /ngrok
!chmod +x /ngrok
```

(範例指令：ch3/cmd2.txt)

```
!mkdir -p /drive/ngrok-ssh
%cd /drive/ngrok-ssh
!wget https://bin.equinox.io/c/4VmDzA7iaHb/ngrok-stable-linux-amd64.zip -O ngrok-stable-linux-amd64.zip
!unzip -u ngrok-stable-linux-amd64.zip
!cp /drive/ngrok-ssh/ngrok /ngrok
!chmod +x /ngrok
```

```
/drive/ngrok-ssh
--2022-03-22 15:47:52--  https://bin.equinox.io/c/4VmDzA7iaHb/ngrok-stable-linux-amd64.zip
Resolving bin.equinox.io (bin.equinox.io)... 52.202.168.65, 54.161.241.46, 54.237.133.81, ...
Connecting to bin.equinox.io (bin.equinox.io)|52.202.168.65|:443... connected.
HTTP request sent, awaiting response... 200 OK
Length: 13832437 (13M) [application/octet-stream]
Saving to: 'ngrok-stable-linux-amd64.zip'

ngrok-stable-linux- 100%[===================>]  13.19M  3.41MB/s    in 6.8s

2022-03-22 15:47:59 (1.94 MB/s) - 'ngrok-stable-linux-amd64.zip' saved [13832437/13832437]

Archive:  ngrok-stable-linux-amd64.zip
```

將自己的 ngrok Authtoken 填入下方程式碼 (替換 <authtoken>)，執行後會將其寫入對應的檔案裡。

```
!/ngrok authtoken <authtoken>
```

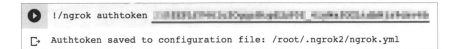

輸入下方程式碼，安裝 flask_ngrok 函式庫。

```
!pip install flask_ngrok
```

```
!pip install flask_ngrok

Collecting flask_ngrok
  Downloading flask_ngrok-0.0.25-py3-none-any.whl (3.1 kB)
Requirement already satisfied: requests in /usr/local/lib/python3.7/di
Requirement already satisfied: Flask>=0.8 in /usr/local/lib/python3.7/
Requirement already satisfied: itsdangerous<2.0,>=0.24 in /usr/local/l
Requirement already satisfied: click<8.0,>=5.1 in /usr/local/pytho
Requirement already satisfied: Jinja2<3.0,>=2.10.1 in /usr/local/lib/p
Requirement already satisfied: Werkzeug<2.0,>=0.15 in /usr/local/lib/p
Requirement already satisfied: MarkupSafe>=0.23 in /usr/local/lib/pyth
Requirement already satisfied: urllib3!=1.25.0,!=1.25.1,<1.26,>=1.21.1
Requirement already satisfied: certifi>=2017.4.17 in /usr/local/lib/py
Requirement already satisfied: idna<3,>=2.5 in /usr/local/lib/python3.
Requirement already satisfied: chardet<4,>=3.0.2 in /usr/local/lib/pyt
Installing collected packages: flask-ngrok
Successfully installed flask-ngrok-0.0.25
```

```
!pip install line-bot-sdk

Collecting line-bot-sdk
  Downloading line_bot_sdk-2.1.0-py2.py3-none-any.whl (83 kB)
     |                                        | 83 kB 2.0 MB
Requirement already satisfied: future in /usr/local/lib/python3.7/dist-pa
Requirement already satisfied: requests>=2.0 in /usr/local/lib/python3.7/
Collecting aiohttp>=3.4
  Downloading aiohttp-3.8.1-cp37-cp37m-manylinux_2_5_x86_64.manylinux1_x8
     |                                        | 1.1 MB 66.9
Requirement already satisfied: typing-extensions>=3.7.4 in /usr/local/lib
Collecting yarl<2.0,>=1.0
  Downloading yarl-1.7.2-cp37-cp37m-manylinux_2_5_x86_64.manylinux1_x86_6
     |                                        | 271 kB 66.5
Requirement already satisfied: attrs>=17.3.0 in /usr/local/lib/python3.7/
Collecting aiosignal>=1.1.2
  Downloading aiosignal-1.2.0-py3-none-any.whl (8.2 kB)
Requirement already satisfied: charset-normalizer<3.0,>=2.0 in /usr/local
Collecting multidict<7.0,>=4.5
  Downloading multidict-6.0.2-cp37-cp37m-manylinux_2_17_x86_64.manylinux2
     |                                        | 94 kB 3.8 MB
Collecting async-timeout<5.0,>=4.0.0a3
  Downloading async_timeout-4.0.2-py3-none-any.whl (5.8 kB)
Collecting asynctest==0.13.0
  Downloading asynctest-0.13.0-py3-none-any.whl (26 kB)
Collecting frozenlist>=1.1.1
  Downloading frozenlist-1.3.0-cp37-cp37m-manylinux_2_5_x86_64.manylinux1
     |                                        | 144 kB 60.5
Requirement already satisfied: idna<3,>=2.5 in /usr/local/lib/python3.7/d
Requirement already satisfied: certifi>=2017.4.17 in /usr/local/lib/pytho
Requirement already satisfied: chardet<4,>=3.0.2 in /usr/local/lib/python
Requirement already satisfied: urllib3!=1.25.0,!=1.25.1,<1.26,>=1.21.1 in
Installing collected packages: multidict, frozenlist, yarl, asynctest, as
Successfully installed aiohttp-3.8.1 aiosignal-1.2.0 async-timeout-4.0.2
```

最後輸入下方的程式碼，內容除了使用 Flask 建立網頁服務，也使用 run_with_ngrok 將網頁服務與 ngrok 串接。

```python
from flask import Flask
from flask_ngrok import run_with_ngrok

app = Flask(__name__)
run_with_ngrok(app)

@app.route("/<name>")
def home(name):
    return f"<h1>hello {name}</h1>"

app.run()
```
（範例程式碼：ch3/code03.py）

執行後，如果出現 ngrok 的網址，就表示已經串接成功，因為 Webhook 通常只支援 https，而 ngrok 也支援 https，屆時複製網址後，需自行將 http 改為 https。

```
 * Serving Flask app "__main__" (lazy loading)
 * Environment: production
   WARNING: This is a development server. Do not use it in a production deployment.
   Use a production WSGI server instead.
 * Debug mode: off
 * Running on http://127.0.0.1:5000/ (Press CTRL+C to quit)
 * Running on http://8747-34-90-116-153.ngrok.io
 * Traffic stats available on http://127.0.0.1:4040
```

串接成功後，就能透過瀏覽器，開啟 ngrok 網址，串連 Colab 所建立的網頁服務。

停止 Colab 的程式，輸入指令安裝 line-bot-sdk 函式庫（pip 前方需要加上！變成 !pip）。

```
!pip install line-bot-sdk
```

安裝完成後，輸入下方的程式碼，**填入將剛剛註冊 LINE BOT 所取得的 LINE Channel access token 和 LINE Channel secret**（後續章節會介紹原理，在此處先依樣畫葫蘆撰寫，確認和 LINE BOT 可以正常串接）。

```python
from flask_ngrok import run_with_ngrok
from flask import Flask, request

# 載入 LINE Message API 相關函式庫
from linebot import LineBotApi, WebhookHandler
from linebot.exceptions import InvalidSignatureError
from linebot.models import MessageEvent, TextMessage, TextSendMessage

# 載入 json 標準函式庫，處理回傳的資料格式
import json

app = Flask(__name__)

@app.route("/", methods=['POST'])
def linebot():
    # 取得收到的訊息內容
    body = request.get_data(as_text=True)
    try:
        # json 格式化訊息內容
        json_data = json.loads(body)
        access_token = '你的 LINE Channel access token'
        secret = '你的 LINE Channel secret'
        # 確認 token 是否正確
        line_bot_api = LineBotApi(access_token)
        # 確認 secret 是否正確
        handler = WebhookHandler(secret)
        # 加入回傳的 headers
        signature = request.headers['X-Line-Signature']
        # 綁定訊息回傳的相關資訊
```

```
        handler.handle(body, signature)
        # 取得 LINE 收到的文字訊息
        msg = json_data['events'][0]['message']['text']
        # 取得回傳訊息的 Token
        tk = json_data['events'][0]['replyToken']
        # 回傳訊息
        line_bot_api.reply_message(tk,TextSendMessage(msg))
        # 印出內容
        print(msg, tk)
    except:
        # 如果發生錯誤，印出收到的內容
        print(body)
    # 驗證 Webhook 使用，不能省略
    return 'OK'
if __name__ == "__main__":
  # 串連 ngrok 服務
  run_with_ngrok(app)
  app.run()
```
(範例程式碼：ch3/code04.py)

點擊 Colab 的執行按鈕，就會得到一串 ngrok 對應的網址，這串網址就是要與 LINE BOT 串接的 Webhook。

```
 * Serving Flask app "__main__" (lazy loading)
 * Environment: production
   WARNING: This is a development server. Do not use it in a production deployment.
   Use a production WSGI server instead.
 * Debug mode: off
 * Running on http://127.0.0.1:5000/ (Press CTRL+C to quit)
 * Running on http://17a0-35-197-40-22.ngrok.io
 * Traffic stats available on http://127.0.0.1:4040
```

複製這串網址後，前往「**LINE Developer 控制台**」，進入自己建立的 LINE BOT Channel，**在 Message API 頁籤裡找到 Webhook setting 選項，貼上 ngrok 網址**（ 注意要使用 https，圖片網址為示意圖，請使用自己 ngrok 產生的網址 ）。

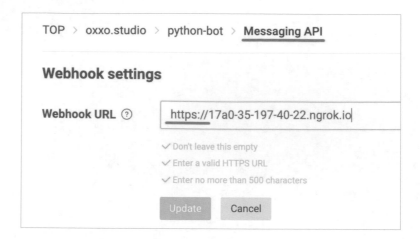

完成後,點擊「Verify」進行驗證,如果**出現 Success 文字表示已經可以正常串接**。

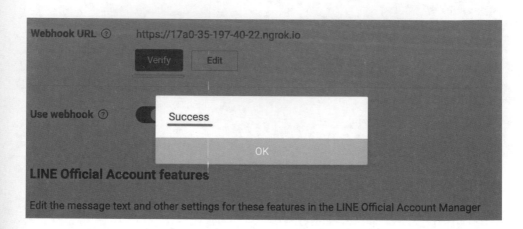

開啟 LINE，找到 LINE BOT 的帳號，傳送訊息給 LINE BOT，LINE BOT 就會
回應相同的訊息。

3-6 建立 Webhook (Google Cloud Functions)

由於使用 Colab + ngrok 所建置的 Webhook，會受限於 Colab 只能運行幾個小時，以及 ngrok 在每次部署都會改變網址的特性，所以無法當作正式的 LINE BOT Webhook (Colab 閒置超過一段時間後還會停止執行並清除安裝的函式庫，需要再次重新安裝)，至於本機環境也會在關機後就失去作用，也無法作為正式的伺服器。

如果要建立一個可以 24 小時不斷運作的 LINE BOT，就可以選擇 Google Cloud Functions 作為 Python 運作的後台，新增並啟用一個 Cloud Functions 程式編輯環境，環境的執行階段選擇 Python (3.7 ～ 3.9 皆可)，進入點改成 linebot (可自訂名稱，之後的程式碼裡也要使用同樣的名稱)。

Google Cloud Functions 詳細教學參考本書第十章。

注意，Google Cloud Functions 只有第一年免費 (Google 贈送 200 美金額度)，超過第一年後，基本每個月使用費約 0.01 美金。

✅ 設定 — ② **程式碼**

執行階段
Python 3.7 ▼ ❓

進入點 *
linebot ❓

建立編輯環境後，點擊 requirements.txt，加入 line-bot-sdk (requirements.txt 的作用是 Cloud Functions 啟用程式時需要安裝的外部函式庫)

點擊 main.py，輸入下方程式碼（因為 Clouds Functions 的 Python 樣板是以 Flask 的基礎建立，所以不需要額外安裝 Flask），**填入將剛剛註冊 LINE BOT 所取得的 LINE Channel access token 和 LINE Channel secret**（後續章節會介紹原理，在此處先依樣畫葫蘆撰寫，確認和 LINE BOT 可以正常串接）。

```python
import json
from linebot import LineBotApi, WebhookHandler
from linebot.exceptions import InvalidSignatureError
from linebot.models import MessageEvent, TextMessage, TextSendMessage

def linebot(request):
    try:
        access_token = '你的 LINE Channel access token'
        secret = '你的 LINE Channel secret'
        body = request.get_data(as_text=True)
        json_data = json.loads(body)
        line_bot_api = LineBotApi(access_token)
        handler = WebhookHandler(secret)
        signature = request.headers['X-Line-Signature']
        handler.handle(body, signature)
        msg = json_data['events'][0]['message']['text']
        tk = json_data['events'][0]['replyToken']
        line_bot_api.reply_message(tk,TextSendMessage(msg))
        print(msg, tk)
    except:
        print(request.args)
    return 'OK'
```
（範例程式碼：ch3/code05.py）

完成後，點擊「部署」，完成後會看見出現「綠色打勾」的圖示（沒有出現綠色打勾表示部署失敗）。

點擊進入程式，選擇「觸發條件」，複製觸發網址（這串網址就是要與 LINE BOT 串接的 Webhook）。

前往「LINE Developer 控制台」，進入自己建立的 LINE BOT Channel，在 Message API 頁籤裡找到 Webhook setting 選項，貼上網址。

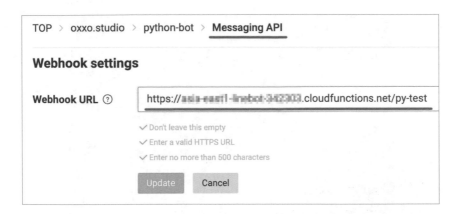

完成後，點擊「Verify」進行驗證，如果**出現 Success 文字表示已經可以正常串接**。

開啟 LINE，找到 LINE BOT 的帳號，傳送訊息給 LINE BOT，LINE BOT 就會回應相同的訊息。

 小結

當這個章節實作完成後，應該已經建立了一個最基本的 LINE BOT（可以發送訊息給 LINE BOT，且 LINE BOT 也會回應同樣的訊息），接下來會繼續介紹更多 LINE BOT 的傳訊息方法，以及主動推播訊息的功能。

4

解析 LINE 訊息

前言

順利將 LINE BOT 串接 Webhook 後，通常會先使用 Flask 函式庫接收訊息，透過 json 標準函式庫將訊息轉換成字典格式，這個章節將會解析各種訊息包含的屬性，理解訊息的內容後，開發 LINE BOT 才能更得心應手。

本章節的範例程式碼：

https://github.com/oxxostudio/book-code/tree/master/linebot/ch4

4-1 將訊息轉換為 json 格式

修改第三章節所建立 Webhook 程式碼（本機環境或 Colab），使用內建的
json 函式庫讀取訊息，並使用 print 印出 json 格式的 body 資料。

```python
# 如果是本機環境不用 flask_ngrok
from flask_ngrok import run_with_ngrok

from flask import Flask, request
import json

app = Flask(__name__)

@app.route("/", methods=['POST'])
def linebot():
    # 讀取資料 json_data
    body = request.get_data(as_text=True)
    # 轉換成 json 格式（字典格式）
    json_data = json.loads(body)
    print(json_data)
    return 'OK'

if __name__ == "__main__":
    # 如果是本機環境不用 run_with_ngrok(app)
    run_with_ngrok(app)
    app.run()
```

（範例程式碼：ch4/code01.py）

完成後執行程式，將連動 ngrok 的 Webhook 填入 LINE Message API 裡，
Verify 驗證通過後，從 LINE 傳送訊息，就能在 Colab 或本機環境的後台顯
示區裡，看見傳送的訊息變成一串 json 的格式，從 json 格式的回應訊息裡，
也能看到許多屬性值（類似下方的格式），接著只要使用字典搭配串列的方
式就能取出指定的內容。

```json
{
  "destination":"Ua3ab05f0fc998ee31756cbc24c4e50f5",
  "events":[
      {
        "type":"message",
        "message":{
            "type":"text",
            "id":"15686965060877",
            "text":"okok"
        },
        "timestamp":1646380955269,
        "source":{
            "type":"user",
            "userId":"XXXXXXXXXXXXXXXXXXXXXXXXXXXXXXX"
        },
        "replyToken":"5df7c3abca3d4956bb04e05d434ea609",
        "mode":"active"
      }
  ]
}
```

通用屬性表示不論哪種格式的訊息,都會出現的屬性值,下方列出 LINE 訊息裡常見的通用屬性:

JSON (JavaScript Object Notation) 是一個資料交換的格式,目前幾乎主流的程式語言解析 JSON 內容,其廣泛使用的程度,使其成為通用的資料格式。

過去在 JSON 尚未普遍前,如果要交換數據資料,往往會透過會寫在 XML 或 TXT 等方式進行交換,然而處理這些檔案格式往往沒有一個統一的標準,然而 JSON 所使用的「物件」格式,可以儲存像是字串、數字、布林值、陣列、物件等各種類型資料格?式,這也是為什麼 JSON 出現之後,馬上就成?為最主流的資料交換格式的原因。

不少線上的服務都有提供美化 JSON 檔案的功能,當 JSON 格式進行整理後,使用者就能更方便觀察資料內容 (例如 https://jsoneditoronline.org/)。

4-2 訊息種類與屬性

不論是接收還是推播訊息，不同種類的訊息都具有不同屬性，接下來會介紹常見的訊息屬性。

🖐 通用屬性

通用屬性表示不論哪種格式的訊息，都會出現的屬性值，下方列出 LINE 訊息裡常見的通用屬性：

屬性	說明	值
destination	回應的目的地。	
events[0].type	訊息狀態。	message（接收訊息），postback（資料訊息），join（加入群組），leave（離開群組）。
events[0].message.type	message 訊息屬性。	text（文字），sticker（表情貼圖），image（圖片），audio（聲音），video（影片），location（位置）。
events[0].message.id	這則訊息的 id。	
events[0].timestamp	傳送訊息的時間。	
events[0].source.type	傳送訊息的來源類型。	user（一對一聊天），group（群組聊天）。
events[0].source.userId	一對一傳送訊息的使用者 ID，如果是群組聊天則會多一個 groupId。	
events[0].replyToken	每則訊息回應使用的 Token。	
events[0].mode	訊息狀態。	active（正常運作中），standby（等待中，不會回應訊息）。

text 文字訊息屬性

文字訊息屬性表示「傳送文字訊息（文字、tag 某人、網址）」時，會出現的訊息屬性。

屬性	說明
events[0].message.text	訊息內容的文字。

sticker 表情貼圖訊息屬性

表情貼圖訊息屬性表示「傳送表情貼圖」時，會出現的訊息屬性。

屬性	說明
events[0].message.stickerId	表情貼圖 ID。
events[0].message.packageId	表情貼圖所在的群組 ID。
events[0].message.stickerResourceType	表情貼圖的類型，STATIC 靜態貼圖，ANIMATION 動態貼圖。
events[0].message.keywords	表情貼圖的關鍵字，以串列表示。

image 圖片訊息屬性

圖片訊息屬性表示傳送「圖片」時，會出現的訊息屬性。

屬性	說明
events[0].message.contentProvider	圖片提供者，line 為使用者提供，external 為程式發送。

video 影片、audio 聲音訊息屬性

如果傳送「影片或聲音」，會出現「時間長度」或「提供者」的訊息屬性。

屬性	說明
events[0].message.duration	影片或聲音的長度，單位毫秒。
events[0].message.contentProvider	影片或聲音提供者，line 為使用者提供，external 為程式發送。

 location 地點位置訊息屬性

如果傳送「地點資訊」，會出現地理位置相關的訊息屬性。

屬性	說明
events[0].message.latitude	緯度。
events[0].message.longitude	經度。
events[0].message.address	地址。

 postback 訊息屬性

如果收到的是「postback 資料」，表示該訊息是直接透過 API 發送，就會出現 data 訊息屬性。

屬性	說明
events[0].postback.data	資料內容。

小結

能夠解析 LINE 的訊息後，就能進一步透過程式判斷該如何回應訊息，例如收到了傷心的貼圖，就可以回覆關心的文字，收到了地址查詢的要求，就能回覆地址的資訊，這也是打造一個 LINE BOT 的必經之路。

Note

5

傳送 LINE 訊息
的方法

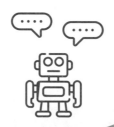

順利將 LINE BOT 串接 Webhook 後，就能開始透過 LINE Message API 實作聊天機器人，接下來的這個章節，會介紹四種傳送 LINE 訊息的方法 (自動回覆訊息、主動推播訊息、使用 Requests 傳送訊息、使用 LINE URL Scheme)，只要掌握這些方法，就能在任何情況下透過 LINE BOT 發送 LINE 訊息。

本章節的範例程式碼：

https://github.com/oxxostudio/book-code/tree/master/linebot/ch5

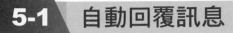

5-1 自動回覆訊息

自動回覆訊息表示「有來有往」的回覆訊息,也就是「收到訊息後才會回覆」,接下來會介紹如何辨識收到的訊息,以及自動回覆指定的訊息。

> 所有程式碼執行前,請參考第三章的開發環境設定,將程式透過 ngrok 產生的 Webhook 網址與 LINE Developer 綁定 (注意,如果是使用 Colab,每次重新執行 ngrok 網址都會變動,需要重新綁定)。

認識 reply token

不論 LINE BOT 接收那種類型的訊息,回覆訊息都是使用 reply_message 方法進行回覆,使用 reply_message 方法在傳送訊息時,必須包含一個 reply token,表示要「回覆哪一個指定的訊息」,**reply token 只會在接收到訊息時產生一次,當再次收到訊息時,就會捨棄前一個 reply token,產生新的 reply token,如果回覆過的訊息需要再次回覆,則「需要再度接收訊息才能回覆」**(如果要多次主動推播訊息,則需使用 push message 的方式處理)。

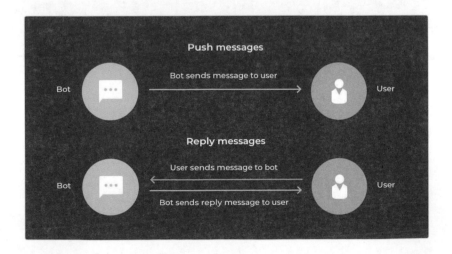

回覆文字訊息

如果要回覆文字訊息,需要先 import TextSendMessage 函式庫,接著就能透過 reply_message 的「**TextSendMessage**」方法,進行文字訊息的回覆,使用方法如下:

```
# 轉換要回傳的訊息
text_message = TextSendMessage(text=msg)

# 回傳訊息,第一個參數為 replyToken,第二個參數為轉換後要回傳的訊息
line_bot_api.reply_message(replyToken,text_message)
```

下方的程式碼執行後,會將收到的訊息轉換成字典格式,並讀取 reply token 和收到的文字,接著就可以使用回傳訊息的方法將同樣的訊息回傳給使用者,不論使用者輸入了什麼文字訊息,LINE BOT 就會回覆一模一樣的文字訊息。

```
# Colab 才需要,本機環境請刪除
from flask_ngrok import run_with_ngrok
from flask import Flask, request
from linebot import LineBotApi, WebhookHandler

# 載入 TextSendMessage 模組
from linebot.models import TextSendMessage
import json

app = Flask(__name__)

@app.route("/", methods=['POST'])
def linebot():
    body = request.get_data(as_text=True)
    json_data = json.loads(body)
    print(json_data)
    try:
        line_bot_api = LineBotApi('你的 Channel access token')
        handler = WebhookHandler(' 你的 LINE Channel secret')
        signature = request.headers['X-Line-Signature']
```

```
        handler.handle(body, signature)
        # 取得 reply token
        tk = json_data['events'][0]['replyToken']
        # 取得使用者發送的訊息
        msg = json_data['events'][0]['message']['text']
        # 設定回傳同樣的訊息
        text_message = TextSendMessage(text=msg)
        # 回傳訊息
        line_bot_api.reply_message(tk,text_message)
    except:
        print('error')
    return 'OK'

if __name__ == "__main__":
    # Colab 才需要，本機環境請刪除
    run_with_ngrok(app)
    app.run()
```

（範例程式碼：ch5/code01.py）

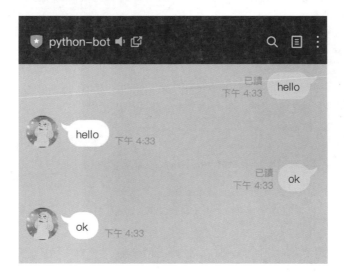

回覆表情貼圖

如果要回覆表情貼圖，需要先 import StickerSendMessage 函式庫，接著就能透過 reply_message 的「**StickerSendMessage**」方法，進行表情貼圖的回覆，使用方法如下：

```
# 設定要回傳的表情貼圖的 stickerId 和 packageId
sticker_message = StickerSendMessage(sticker_id=stickerId, package_id=packageId)

# 回傳訊息，第一個參數為 replyToken，第二個參數為轉換後要回傳的訊息
line_bot_api.reply_message(replyToken,sticker_message)
```

stickerId 和 packageId 分別代表「表情貼圖所在群組 ID」和「表情貼圖 ID」，由於 LINE BOT 預設的表情貼圖只有 LINE 官方所提供的貼圖，可以透過 List of available stickers (https://developers.line.biz/en/docs/messaging-api/sticker-list/) 網站查詢 ID 號碼。

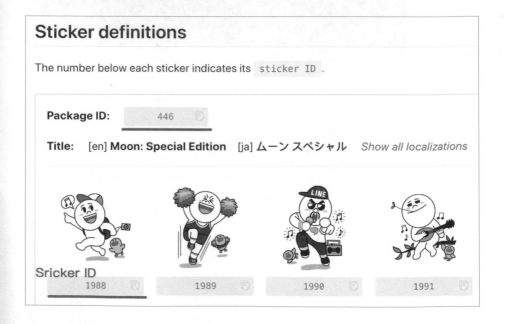

下方的程式碼執行後，當使用者傳送了表情貼圖，會將收到的字典格式訊息裡，讀取 reply token 和收到的表情 ID，LINE BOT 就會回覆一模一樣的表情貼圖 (限制為官方預設的表情貼圖)。

```python
# Colab 才需要，本機環境請刪除
from flask_ngrok import run_with_ngrok
from flask import Flask, request
from linebot import LineBotApi, WebhookHandler

# 載入 StickerSendMessage 模組
from linebot.models import StickerSendMessage
import json

app = Flask(__name__)

@app.route("/", methods=['POST'])
def linebot():
    body = request.get_data(as_text=True)
    json_data = json.loads(body)
    print(json_data)
    try:
        line_bot_api = LineBotApi(' 你的 Channel access token')
        handler = WebhookHandler(' 你的 LINE Channel secret')
        signature = request.headers['X-Line-Signature']
        handler.handle(body, signature)
        # 取得 reply token
        tk = json_data['events'][0]['replyToken']
        # 取得 stickerId
        stickerId = json_data['events'][0]['message']['stickerId']
        # 取得 packageId
        packageId = json_data['events'][0]['message']['packageId']
        # 設定要回傳的表情貼圖
        sticker_message = StickerSendMessage(sticker_id=stickerId,
package_id=packageId)
        # 回傳訊息
        line_bot_api.reply_message(tk,sticker_message)
    except:
        print('error')
```

```
    return 'OK'

if __name__ == "__main__":
    # Colab 才需要，本機環境請刪除
    run_with_ngrok(app)
    app.run()
```
（範例程式碼：ch5/code02.py）

回覆圖片或影片訊息

import ImageSendMessage 或 VideoSendMessage 函式庫之後，就能透過 reply_message 的「**ImageSendMessage**」或「**VideoSendMessage**」方法，回覆圖片或影片，使用方法如下：

```
# 回傳圖片
# 設定要回傳的 original_content_url 原始網址、preview_image_url 縮圖網址
img_message = ImageSendMessage(original_content_url=img_url, preview_
image_url=img_url)

# 回傳訊息，第一個參數為 replyToken，第二個參數為轉換後要回傳的訊息
line_bot_api.reply_message(replyToken,img_message)


# 回傳影片
# 設定要回傳的 original_content_url 原始網址、preview_image_url 縮圖網址
img_message = VideoSendMessage(original_content_url=img_url, preview_
image_url=img_url)

# 回傳訊息，第一個參數為 replyToken，第二個參數為轉換後要回傳的訊息
line_bot_api.reply_message(replyToken,img_message)
```

下方的程式碼執行後，當使用者傳送了某個文字，LINE BOT 就會回覆跟這個文字有關的圖片 (先建立好圖片網址和文字對照的字典，範例圖片來源是維基百科)，如果找不到文字對應的圖片，就會回傳「找不到相關圖片」。

```
# Colab 才需要，本機環境請刪除
from flask_ngrok import run_with_ngrok
from flask import Flask, request
from linebot import LineBotApi, WebhookHandler
# 載入 TextSendMessage 和 ImageSendMessage 模組
from linebot.models import TextSendMessage, ImageSendMessage
import json

app = Flask(__name__)

@app.route("/", methods=['POST'])
def linebot():
    body = request.get_data(as_text=True)
    json_data = json.loads(body)
    print(json_data)
    try:
        line_bot_api = LineBotApi(' 你的 Channel access token')
        handler = WebhookHandler(' 你的 LINE Channel secret')
```

```python
        signature = request.headers['X-Line-Signature']
        handler.handle(body, signature)
        tk = json_data['events'][0]['replyToken']
        msg = json_data['events'][0]['message']['text']
        # 取得對應的圖片，如果沒有取得，會是 False
        img_url = reply_img(msg)
        if img_url:
            # 如果有圖片網址，回傳圖片
            img_message = ImageSendMessage(original_content_url=img_url, preview_image_url=img_url)
            line_bot_api.reply_message(tk,img_message)
        else:
            # 如果是 False，回傳文字
            text_message = TextSendMessage(text=' 找不到相關圖片 ')
            line_bot_api.reply_message(tk,text_message)
    except:
        print('error')
    return 'OK'
# 建立回覆圖片的函式
def reply_img(text):
    # 文字對應圖片網址的字典
    img = {
        ' 皮卡丘 ':'https://upload.wikimedia.org/wikipedia/en/a/a6/Pok%C3%A9mon_Pikachu_art.png',
        ' 傑尼龜 ':'https://upload.wikimedia.org/wikipedia/en/5/59/Pok%C3%A9mon_Squirtle_art.png'
    }
    if text in img:
      return img[text]
    else:
      # 如果找不到對應的圖片，回傳 False
      return False

if __name__ == "__main__":
    # Colab 才需要，本機環境請刪除
    run_with_ngrok(app)
    app.run()
```

(範例程式碼：ch5/code03.py)

回覆地址訊息

import LocationSendMessage 函式庫之後，就能透過 reply_message 裡的「**LocationSendMessage**」方法，回覆地址訊息，使用方法如下：

```
# 設定要回傳的 title 地圖標題
# address 地址標示、latitude 緯度、longitude 經度
location_message = LocationSendMessage(title=location_dect['title'],
                        address=location_dect['address'],
                        latitude=location_dect['latitude'],
                        longitude=location_dect['longitude'])

# 回傳訊息，第一個參數為 replyToken，第二個參數為轉換後要回傳的訊息
line_bot_api.reply_message(replyToken,location_message)
```

下方的程式碼執行後，當使用者傳送了某個地點的文字，LINE BOT 就會回覆跟這個地點的地圖（先建立文字和地址、經緯度的對照字典），如果找不到文字對應的地址，就會回傳「找不到相關地點」。

```python
# Colab 才需要，本機環境請刪除
from flask_ngrok import run_with_ngrok
from flask import Flask, request
from linebot import LineBotApi, WebhookHandler
# 載入 TextSendMessage 和 LocationSendMessage 模組
from linebot.models import TextSendMessage, LocationSendMessage
import json

app = Flask(__name__)

@app.route("/", methods=['POST'])
def linebot():
    body = request.get_data(as_text=True)
    json_data = json.loads(body)
    print(json_data)
    try:
        line_bot_api = LineBotApi('你的 Channel access token')
        handler = WebhookHandler('你的 LINE Channel secret')
        signature = request.headers['X-Line-Signature']
        handler.handle(body, signature)
        tk = json_data['events'][0]['replyToken']
        msg = json_data['events'][0]['message']['text']
        # 取得對應的地址，如果沒有取得，會是 False
        location_dect = reply_location(msg)
        if location_dect:
            # 如果有地點資訊，回傳地點
            location_message = LocationSendMessage(title=location_
dect['title'],
                                address=location_dect['address'],
                                latitude=location_dect['latitude'],
longitude=location_dect['longitude'])
            line_bot_api.reply_message(tk,location_message)
```

```
        else:
            # 如果是 False，回傳文字
            text_message = TextSendMessage(text=' 找不到相關地點 ')
            line_bot_api.reply_message(tk,text_message)
    except:
        print('error')
    return 'OK'
# 建立回覆地點的函式
def reply_location(text):
    # 建立地點與文字對應的字典
    location = {
        '101':{
            'title':' 台北 101',
            'address':'110 台北市信義區信義路五段 7 號 ',
            'latitude':'25.034095712145003',
            'longitude':'121.56489941996108'
        },
        ' 總統府 ':{
            'title':' 總統府 ',
            'address':'100 台北市中正區重慶南路一段 122 號 ',
            'latitude':'25.0403198747750914',
            'longitude':'121.51162883484746'
        }
    }
    if text in location:
      return location[text]
    else:
      # 如果找不到對應的地點，回傳 False
      return False


if __name__ == "__main__":
    # Colab 才需要，本機環境請刪除
    run_with_ngrok(app)
    app.run()
```

(範例程式碼：ch5/code04.py)

👆 Google Cloud Functions 部署自動回覆訊息

因為使用 ngrok + Colab 的 Python 程式，只會運作幾個小時就停止，甚至再次執行時需要重新安裝相關函式庫，而本機環境也會在關機或更換 IP 後失去其作用，所以都只能作為「開發中」使用，如果要真正建構 LINE BOT 的 Python 程式，可以使用 Google Cloud Functions 部署程式。

參考本書第三章「建立 Webhook (Google Cloud Functions)」一節，建立好 Cloud Functions 的環境後，整合上述的回覆方式，部署下方的程式碼 (記得輸入自己的 Access Token 和 Channel secret)。

```python
import json
from linebot import LineBotApi, WebhookHandler
# 載入對應的函式庫
from linebot.models import TextSendMessage, StickerSendMessage,
ImageSendMessage, LocationSendMessage


def linebot(request):
    try:
        body = request.get_data(as_text=True)
        # json 格式化收到的訊息
        json_data = json.loads(body)
        line_bot_api = LineBotApi(' 你的 Channel access token')
        handler = WebhookHandler(' 你的 Channel secret')
        signature = request.headers['X-Line-Signature']
        handler.handle(body, signature)
        # 取得 reply token
        tk = json_data['events'][0]['replyToken']
        # 取得 message 的類型
        tp = json_data['events'][0]['message']['type']
        if tp == 'text':
            # 如果是文字類型的訊息，取出文字並對應到 reply_msg 的函式
            msg = reply_msg(json_data['events'][0]['message']['text'])
            if msg[0] == 'text':
                # 如果要回傳的訊息是 text，使用 TextSendMessage 方法
                line_bot_api.reply_message(tk,TextSendMessage
(text=msg[1]))
            if msg[0] == 'location':
                # 如果要回傳的訊息是 location，使用 LocationSendMessage 方法
                line_bot_api.reply_message(tk,LocationSendMessage
(title=msg[1]['title'],
address=msg[1]['address'],
latitude=msg[1]['latitude'],
longitude=msg[1]['longitude']))
            if msg[0] == 'image':
                # 如果要回傳的訊息是 image，使用 ImageSendMessage 方法
                line_bot_api.reply_message(tk,ImageSendMessage
(original_content_url=msg[1],
preview_image_url=msg[1]))
        if tp == 'sticker':
```

```
                # 如果收到的訊息是表情貼圖
                stickerId = json_data['events'][0]['message']['stickerId']
# 取得 stickerId
                packageId = json_data['events'][0]['message']['packageId']
# 取得 packageId
                # 使用 StickerSendMessage 方法回傳同樣的表情貼圖
                line_bot_api.reply_message(tk,StickerSendMessage
(sticker_id=stickerId, package_id=packageId))
            if tp == 'location':
                # 如果是收到的訊息是地點資訊
                line_bot_api.reply_message(tk,TextSendMessage
(text=' 好地點！'))
            if tp == 'image':
                # 如果是收到的訊息是圖片
                line_bot_api.reply_message(tk,TextSendMessage
(text=' 好圖給讚！'))
            if tp == 'audio':
                # 如果是收到的訊息是聲音
                line_bot_api.reply_message(tk,TextSendMessage
(text=' 聲音讚喔～'))
            if tp == 'video':
                # 如果是收到的訊息是影片
                line_bot_api.reply_message(tk,TextSendMessage
(text=' 影片內容真是不錯！'))
    except:
        print('error', body)
    return 'OK'
# 定義回覆訊息的函式
def reply_msg(text):
    # 客製化回覆文字
    msg_dict = {
        'hi':'Hi! 你好呀～',
        'hello':'Hello World!!!!',
        ' 你好 ':' 你好呦～',
        'help':' 有什麼要幫忙的嗎？'
    }
    # 如果出現特定地點，提供地點資訊
    local_dict = {
        ' 總統府 ':{
```

```
            'title':'總統府',
            'address':'100 台北市中正區重慶南路一段 122 號',
            'latitude':'25.040319874750914',
            'longitude':'121.51162883484746'
        }
    }
    # 如果出現特定圖片文字，提供圖片網址
    img_dict = {
        '皮卡丘':'https://upload.wikimedia.org/wikipedia/en/a/a6/
Pok%C3%A9mon_Pikachu_art.png',
        '傑尼龜':'https://upload.wikimedia.org/wikipedia/en/5/59/
Pok%C3%A9mon_Squirtle_art.png'
    }
    # 預設回覆的文字就是收到的訊息
    reply_msg_content = ['text',text]
    if text in msg_dict:
        reply_msg_content = ['text',msg_dict[text.lower()]]
    if text in local_dict:
        reply_msg_content = ['location',local_dict[text.lower()]]
    if text in img_dict:
        reply_msg_content = ['image',img_dict[text.lower()]]
    return reply_msg_content
```

（範例程式碼：ch5/code05.py）

部署成功後 (出現綠色打勾圖示)，更新 LINE Developer 的 Webhook，驗證 Webhook 沒問題後，就可以在 LINE 與自己開發的 LINE BOT 聊天了，不過回覆訊息的機制是「一來一往」，只有在發送訊息後取得 reply token 才能回覆訊息，如果要主動推播訊息，則需要使用 push messsage 的方式才能實現。

5-2　主動推播訊息

有別於 LINE BOT 自動回覆訊息，主動推播訊息表示可以在特定事件被觸發時，主動傳送訊息到使用者端，不過 LINE 的主動推播訊息只提供開發者免費 50 個好友額度，如果是官方帳號具有大量好友的推播訊息需求，就需要額外付費。

> 所有程式碼執行前，請參考第三章的開發環境設定，將程式透過 ngrok 產生的 Webhook 網址與 LINE Developer 綁定 (注意，如果是使用 Colab，每次重新執行 ngrok 網址都會變動，需要重新綁定)。

🖐 取得 LINE user ID

回顧第二章「建立 LINE BOT」，進入 LINE Developers 的 LINE Channel，在 **Basic Setting** 裡找到 user ID。

TOP ＞ oxxo.studio ＞ python-bot ＞ **Basic settings**

Your user ID ⑦　　　U831328b951316aa0a50c8D003b1f638055

在 Message API 裡找到 Channel Access Token。

TOP ＞ oxxo.studio ＞ python-bot ＞ **Messaging API**

Channel access token

Channel access token (long-lived) ⑦

iKifuZV0OyREh6AbZnfhwHv75tUgkq3KSN134Dq9rr2xQhxrrF8fHr+32GY0A90EMXxxU+YTtabpfoeBdUywf R92YF1TqNUMuk25/1CaNS=3NFMUXB9hDYYqf3/o54CQUw3rewCFPWL2kjDb6MeTzhwd804t99U10/w1 cDryifU=

推播訊息的方法

推播訊息使用 **line_bot_api.push_message()** 方法，訊息可以是文字、表情貼圖、圖片、地點 ... 等類型，每種類型都有對應的推播訊息方法 (需要個別 import 函式庫，基本上和自動回覆訊息相同)，下方列出常用類型的推播方法：

文字類型

```
# 轉換要推播的訊息
text_message = TextSendMessage(text=msg)

# 推播訊息，第一個參數為 replyToken，第二個參數為轉換後要回傳的訊息
line_bot_api.push_message(replyToken,text_message)
```

表情貼圖

```
# 設定要推播的表情貼圖的 stickerId 和 packageId
sticker_message = StickerSendMessage(sticker_id=stickerId, package_id=packageId)

# 推播訊息，第一個參數為 replyToken，第二個參數為轉換後要回傳的訊息
line_bot_api.push_message(replyToken,sticker_message)
```

圖片或影片

```
# 推播圖片
# 設定要推播的 original_content_url 原始網址、preview_image_url 縮圖網址
img_message = ImageSendMessage(original_content_url=img_url, preview_image_url=img_url)

# 推播訊息，第一個參數為 replyToken，第二個參數為轉換後要回傳的訊息
line_bot_api.push_message(replyToken,img_message)


# 推播影片
# 設定要推播的 original_content_url 原始網址、preview_image_url 縮圖網址
img_message = VideoSendMessage(original_content_url=img_url, preview_
```

```
image_url=img_url)

# 推播訊息，第一個參數為 replyToken，第二個參數為轉換後要回傳的訊息
line_bot_api.push_message(replyToken,img_message)
```

地圖

```
# 設定要推播的 title 地圖標題
# address 地址標示、latitude 緯度、longitude 經度
location_message = LocationSendMessage(title=location_dect['title'],
                                       address=location_
dect['address'],
                                       latitude=location_
dect['latitude'],
                                       longitude=location_
dect['longitude'])

# 推播訊息，第一個參數為 replyToken，第二個參數為轉換後要回傳的訊息
line_bot_api.push_message(replyToken,location_message)
```

在本機環境或 Colab 撰寫程式，填入自己的 LINE User ID 和 Access Token，程式會 **使用 Flask 的 request 取得網址的 msg 參數**，判斷如果有 **msg 參數**，就使用 **LINE Message API** 的 **push_message**，讓 **LINE BOT 主動推播 msg 參數的值**。

```
# Colab 才需要，本機環境請刪除
from flask_ngrok import run_with_ngrok
from flask import Flask, request
from linebot import LineBotApi, WebhookHandler
from linebot.models import TextSendMessage

app = Flask(__name__)

@app.route("/")
def home():
  line_bot_api = LineBotApi('你的 access token')
  try:
    # 取得網址的 msg 參數
```

```
    msg = request.args.get('msg')
    if msg != None:
      # 如果有 msg 參數，觸發 LINE Message API 的 push_message 方法
      line_bot_api.push_message(' 你的 User ID',
TextSendMessage(text=msg))
      return msg
    else:
      return 'OK'
  except:
    print('error')

if __name__ == "__main__":
    # Colab 才需要，本機環境請刪除
    run_with_ngrok(app)
    app.run()
```
（範例程式碼：ch5/code06.py）

完成後，執行程式，從瀏覽器開啟 ngrok 產生的網址，**在網址後方輸入「?msg=XXXX」**，重新整理後，如果有加入這個 LINE BOT 為好友，就會收到 LINE BOT 推播的訊息。

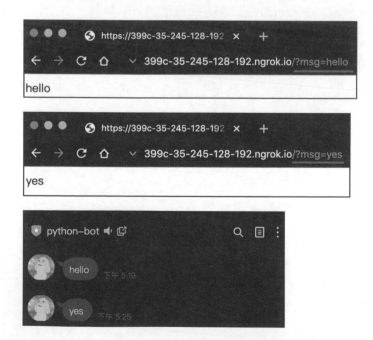

修改程式，加入額外的邏輯判斷，讓判斷不同訊息的數字時，就會主動推播
不同格式的訊息。

```python
# Colab 才需要，本機環境請刪除
from flask_ngrok import run_with_ngrok
from flask import Flask, request
from linebot import LineBotApi, WebhookHandler
# 載入對應的函式庫
from linebot.models import TextSendMessage, StickerSendMessage,
ImageSendMessage, LocationSendMessage
app = Flask(__name__)

@app.route("/")
def home():
  line_bot_api = LineBotApi('你的 access token')
  try:
    msg = request.args.get('msg')
    if msg == '1':
      # 如果 msg 等於 1，發送文字訊息
      line_bot_api.push_message('你的 user ID',
TextSendMessage(text='hello'))
    elif msg == '2':
      # 如果 msg 等於 2，發送表情貼圖
      line_bot_api.push_message('你的 user ID',
StickerSendMessage(package_id=1, sticker_id=2))
    elif msg == '3':
      # 如果 msg 等於 3，發送圖片
      imgurl = 'https://upload.wikimedia.org/wikipedia/en/a/a6/
Pok%C3%A9mon_Pikachu_art.png'
      line_bot_api.push_message('你的 user ID',
ImageSendMessage(original_content_url=imgurl, preview_image_
url=imgurl))
    elif msg == '4':
      # 如果 msg 等於 4，發送地址資訊
      line_bot_api.push_message('你的 user ID',
LocationSendMessage(title='總統府',
                                address='100台北市中正區重慶南路一段122號',
                                latitude='25.040319874750914',
                                longitu de='121.51162883484746'))
```

```
    else:
      msg = 'ok'    # 如果沒有 msg 或 msg 不是 1～4，將 msg 設定為 ok
    return msg
  except:
    print('error')



if __name__ == "__main__":
    # Colab 才需要，本機環境請刪除
    run_with_ngrok(app)
    app.run()
```
(範例程式碼：ch5/code07.py)

程式執行後，從瀏覽器開啟 ngrok 產生的網址，修改後方 msg 的內容，例如 msg=1 時就會發送文字，msg=2 就會發送貼圖。

Google Cloud Functions 部署推播訊息

因為使用 ngrok + Colab 的 Python 程式，只會運作幾個小時就停止，甚至再次執行時需要重新安裝相關函式庫，而本機環境也會在關機或更換 IP 後失去其作用，所以都只能作為「開發中」使用，如果要真正建構 LINE BOT 的 Python 程式，可以使用 Google Cloud Functions 部署程式。

參考本書第三章「建立 Webhook (Google Cloud Functions)」一節，建立好 Cloud Functions 的環境後，整合上述的推播方式，部署下方的程式碼 (記得輸入自己的 Access Token 和 user ID)。

```
from linebot import LineBotApi, WebhookHandler
from linebot.models import TextSendMessage, StickerSendMessage,
ImageSendMessage, LocationSendMessage

def pushmsg(request):
  line_bot_api = LineBotApi('你的 access token')
  try:
    msg = request.args.get('msg')
    if msg == '1':
      line_bot_api.push_message('你的 user ID',
TextSendMessage(text='hello'))
    elif msg == '2':
      line_bot_api.push_message('你的 user ID',
StickerSendMessage(package_id=1, sticker_id=2))
    elif msg == '3':
      imgurl = 'https://upload.wikimedia.org/wikipedia/en/a/a6/
Pok%C3%A9mon_Pikachu_art.png'
      line_bot_api.push_message('你的 user ID',
ImageSendMessage(original_content_url=imgurl, preview_image_
url=imgurl))
    elif msg == '4':
      line_bot_api.push_message('你的 user ID',
LocationSendMessage(title='總統府',
                            address='100台北市中正區重慶南路一段122號',
                            latitude='25.040319874750914',
                            longitude='121.51162883484746'))
```

```
    else:
      msg = 'ok'
    return msg
  except:
    print('error')
```
(範例程式碼：ch5/code08.py)

部署成功後 (出現綠色打勾圖示)，更新 LINE Developer 的 Webhook，驗證 Webhook 沒問題後，就可以開啟瀏覽器，輸入 Cloud Functions 裡觸發的網址，修改 msg 的值後就會主動推播不同訊息。透過主動推播訊息的方法，就能在特定事件被觸發時，讓 LINE BOT 發送訊息通知，或進一步搭配 reply message 的方法，做出更進階的 LINE BOT。

5-3 使用 Requests 傳送訊息

除了可以透過 LINE Message API 傳送訊息，LINE 也提供使用 requests 的方式傳送訊息，只要輸入正確的 Access Token，就能在任何地方與 LINE 溝通，甚至可以發送樣板按鈕、Flex Message... 等特殊訊息 (後續章節會介紹)，接下來會介紹如何使用 requests 的方式，直接從後端伺服器傳送訊息。

> requests 函式庫教學：
>
> https://steam.oxxostudio.tw/category/python/spider/requests.html

reply message 回覆訊息

使用 requests 函式庫 (本機環境可能要額外安裝)，搭配 LINE reply message 的 API，輸入收到訊息後的 replyToken，就能向發送訊息的使用者傳送訊息。

- replyToken 只有在收到使用者發送訊息時才會包含在訊息裡，可以參考本章的「自動回覆訊息」
- LINE reply message 的 API 為：https://api.line.me/v2/bot/message/reply

```python
import requests, json

# 設定 request 的 headers，注意前方要有 Bearer
headers = {'Authorization':'Bearer 你的 access token','Content-Type':'application/json'}

# 設定 request 的 body，必須包含 replyToken 和 messages
body = {
    'replyToken':replyToken,
    'messages':[{
```

```
            'type': 'text',
            'text': 'hello'
        }]
}

# 使用 POST 方法發出請求
req = requests.request('POST', 'https://api.line.me/v2/bot/message/
reply', headers=headers,data=json.dumps(body).encode('utf-8'))
print(req.text)
```
(範例程式碼：ch5/code09.py)

👆 push message 主動傳送訊息

使用 requests 函式庫 (本機環境可能要額外安裝)，搭配 LINE push message 的 API，輸入指定使用者的 user ID，執行程式後，就能向使用者發送訊息，透過 Requests 的方法，就能讓 LINE 的用途更加廣泛。

> 使用者的 user ID 除了可以從後台看到開發者 ID，也會包含在使用者傳送的訊息中。
>
> ● LINE push message 的 API 為：
>
>   https://api.line.me/v2/bot/message/push

```
import requests, json

# 設定 request 的 headers，注意前方要有 Bearer
headers = {'Authorization':'Bearer 你的 access token','Content-
Type':'application/json'}

# 設定 request 的 body，必須包含 to 和 messages
body = {
    'to':'你的 user ID',
    'messages':[{
            'type': 'text',
            'text': 'hello'
```

```
        }]
    }
# 向指定網址發送 request
req = requests.request('POST', 'https://api.line.me/v2/bot/message/
push',headers=headers,data=json.dumps(body).encode('utf-8'))
# 印出得到的結果
print(req.text)
```
(範例程式碼：ch5/code10.py)

透過 API 所傳送的訊息格式

使用 API 的方式傳送訊息，必須要符合各個類型的格式，下方列出相關的訊
息格式：

● 文字訊息格式

```
{
    "type": "text",
    "text": "Hello, world"
}
```

● 表情貼圖格式（ 表情 ID 參考：https://developers.line.biz/en/docs/
messaging-api/sticker-list/#specify-sticker-in-message-object ）

```
{
    "type": "sticker",
    "packageId": "446",
    "stickerId": "1988"
}
```

● 圖片訊息格式（ originalContentUrl 原始圖片網址，previewImageUrl 預
覽圖片網址 ）

```
{
    "type": "image",
    "originalContentUrl": "https://example.com/original.jpg",
    "previewImageUrl": "https://example.com/preview.jpg"
```

```
}
```

● 影片訊息格式（**originalContentUrl** 影片網址，**previewImageUrl** 預覽圖片網址）

```
{
    "type": "video",
    "originalContentUrl": "https://example.com/original.mp4",
    "previewImageUrl": "https://example.com/preview.jpg",
    "trackingId": "track-id"
}
```

● 聲音訊息格式（**originalContentUrl** 聲音網址，**duration** 聲音長度）

```
{
    "type": "audio",
    "originalContentUrl": "https://example.com/original.m4a",
    "duration": 60000
}
```

● 地點位置訊息格式（**title** 地點名稱，**address** 地址，**latitude** 緯度，**longitude** 經度）

```
{
    "type": "location",
    "title": "my location",
    "address": "1-6-1 Yotsuya, Shinjuku-ku, Tokyo, 160-0004, Japan",
    "latitude": 35.687574,
    "longitude": 139.72922
}
```

5-4 使用 LINE URL Scheme

雖然 LINE Message API 已經具有許多好用的方法，可以觸發發送訊息、地點或呼叫相機 ... 等行為事件，但其實 LINE 也提供 URL Scheme 的方式，只要透過特定的網址和參數，就能實現和 Message API 類似的功能，接下來將會介紹如何使用 LINE URL Scheme。

認識 LINE URL Scheme

LINE URL Scheme 是一個「網址」，**當使用者點擊該網址，或透過 LINE 的 API 執行網址，就會根據網址的參數，要求 LINE 進行對應的動作**（目前 LINE URL Scheme 只支援行動裝置的 LINE，不支援 windows 或 MccOS 桌面版）。

LINE URL Scheme 的網址格式如下：

```
https://line.me/R/ 動作
```

發送訊息

開啟或點擊下方的網址，會開啟 LINE，詢問是否要發送指定的訊息，選擇要傳送的聯絡人，就可以發送訊息。

```
https://line.me/R/share?text={ 要發送的訊息 }
```

開啟或點擊下方的網址，會開啟指定的 LINE 官方帳號聊天畫面，並將網址內的訊息帶入對話輸入框裡。

```
https://line.me/R/oaMessage/@ 官方帳號 id/ 要發送的訊息
```

發送地點資訊

開啟或點擊下方的網址,就會在 LINE 裡開啟地圖定位的功能,指定地點按下分享,就會將地點分享出去。

```
https://line.me/R/nv/location/
```

開啟相機、發送圖片

開啟或點擊下方的網址,會呼叫 LINE 打開相機功能,拍照後將圖片傳送出去。

```
https://line.me/R/nv/camera/
```

開啟或點擊下方的網址,會透過 LINE 選擇手機的相簿,從相簿中開啟單一張圖片或多張圖片。

https://line.me/R/nv/cameraRoll/single	選擇一張圖片進行分享。
https://line.me/R/nv/cameraRoll/multi	選擇多張圖片進行分享。

常用功能列表

下方列出其他透過 LINE URL Scheme 可以實現的常用功能（完整 LINE URL Scheme 參考：https://developers.line.biz/en/docs/line-login/using-line-url-scheme/#available-line-url-schemes ）

LINE URL Scheme	說明
https://line.me/R/nv/profile	開啟個人資料畫面。
https://line.me/R/nv/profileSetId	開啟個人 ID 畫面。
https://line.me/R/nv/chat	開啟聊天畫面。
https://line.me/R/nv/timeline	開啟 LINE VOOM 畫面。
https://line.me/R/nv/wallet	開啟錢包畫面。
https://line.me/R/nv/addFriends	開啟新增朋友畫面。
https://line.me/R/nv/officialAccounts	開啟官方帳號畫面。
https://line.me/R/nv/settings	開啟設定畫面。
https://line.me/R/nv/settings/sticker	開啟貼圖畫面。

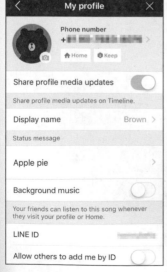
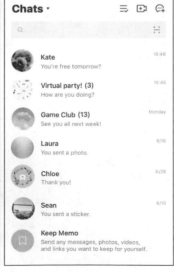
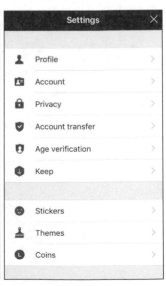

 小結

透過這個章節介紹的四種傳送 LINE 訊息的方法（自動回覆訊息、主動推播訊息、使用 Requests 傳送訊息、使用 LINE URL Scheme），就能讓傳送訊息變得更加靈活彈性，在任何情況下都能透過 LINE BOT 發送 LINE 訊息。

Note

Note

6

傳送不同類型
的 LINE 訊息

前言

了解基本四種傳送 LINE 訊息的方法之後，就可以開始嘗試發送一些比較不同的訊息，例如樣板訊息、Flex Message，或者替 LINE BOT 建立圖文選單（在聊天視窗下方別具特色的選單），接下來的這個章節，將會介紹相關的操作方法。

本章節的範例程式碼：

https://github.com/oxxostudio/book-code/tree/master/linebot/ch6

6-1 發送樣板訊息

LINE BOT 除了單純的發送文字與表情訊息,也可傳送「樣板訊息 template message」給使用者,讓使用者可以直接在上面選擇或進行確認,熟悉樣板訊息後,就能夠提供使用者可以點選的「選項」,快速讓使用者選擇,避免還要額外撰寫程式從對話語句判斷關鍵字,是相當方便好用的功能。(注意,樣板訊息只能在行動裝置上使用)

ButtonsTemplate 按鈕樣板

ButtonsTemplate 按鈕樣板會以聊天對話框的方式出現,在對話框裡提供最多「四個按鈕」,供使用者點選,相關屬性說明如下表所示:

屬性	說明
thumbnailImageUrl	縮圖連結,支援 jpg 和 png,最大寬度 1024px。
imageAspectRatio	圖片比例,預設 rectangle (1.51:1),可設定 square (1:1)。
imageSize	圖片尺寸,預設 cover (絕對撐滿,才切超過顯示比例的區域),可設定 contain。
imageBackgroundColor	放置圖片區域的背景顏色,預設白色 #FFFFFF。
title	樣板標題。
text	樣板說明文字。
actions	點擊按鈕觸發的行為,一個按鈕一種行為,最多支援四個按鈕。

import TemplateSendMessage 和 ButtonsTemplate 函式庫之後,就能透過 ButtonsTemplate() 方法發送按鈕樣板訊息,**按鈕樣板的按鈕是透過 actions 屬性設定,actions 屬性為 list 格式,長度最多為 4,內容可以是 PostbackAction (發送 postback)、MessageAction (發送文字) 或 URIAction (前往超連結)。**

下方的程式碼範例中，使用 line_bot_api.push_message() 方法，搭配 ButtonsTemplate() 方法發送按鈕樣板。

```
from linebot import LineBotApi, WebhookHandler
# 需要額外載入對應的函示庫
from linebot.models import PostbackAction,URIAction, MessageAction,
TemplateSendMessage, ButtonsTemplate
line_bot_api = LineBotApi('你的 Channel access token')
line_bot_api.push_message('你的 user ID', TemplateSendMessage(
    alt_text='ButtonsTemplate',
    template=ButtonsTemplate(
     thumbnail_image_url='https://steam.oxxostudio.tw/download/python/
line-template-message-demo.jpg',
        title='OXXO.STUDIO',
        text='這是按鈕樣板',
        actions=[
            PostbackAction(
                label='postback',
                data='發送 postback'
            ),
            MessageAction(
                label='說 hello',
                text='hello'
            ),
            URIAction(
                label='前往 STEAM 教育學習網',
                uri='https://steam.oxxostudio.tw'
            )
        ]
    )
))
```

（範例程式碼：ch6/code01.py）

程式執行後，就會看見 LINE BOT 向自己發送三顆按鈕的按鈕樣板訊息。

ConfirmTemplate 確認樣板

ConfirmTemplate 確認樣板會提供「兩個按鈕」供使用者選擇，相關屬性如下：

屬性	說明
text	樣板説明文字。
actions	點擊按鈕觸發的行為，一個按鈕一種行為，最多支援兩個按鈕。

import TemplateSendMessage 和 ConfirmTemplate 函式庫之後，就能透過 ConfirmTemplate() 方法發送確認樣板訊息，**確認樣板的按鈕是透過 actions 屬性設定，actions 屬性為 list 格式，長度最多為 2，內容是 MessageAction (文字)。**

下方的程式碼範例中,使用 line_bot_api.push_message() 方法,搭配 ConfirmTemplate() 方法發送確認樣板。

```python
from linebot import LineBotApi, WebhookHandler
# 需要額外載入對應的函示庫
from linebot.models import MessageAction, TemplateSendMessage,
ConfirmTemplate
line_bot_api = LineBotApi('你的 Channel access token')
line_bot_api.push_message('你的 user ID', TemplateSendMessage(
    alt_text='ConfirmTemplate',
    template=ConfirmTemplate(
            text='你好嗎?',
            actions=[
                MessageAction(
                    label='好喔',
                    text='好喔'
                ),
                MessageAction(
                    label='好喔',
                    text='不好喔'
                )
            ]
        )
))
```

(範例程式碼:ch6/code02.py)

程式執行後,就會看見 LINE BOT 向自己發送兩顆按鈕的確認樣板訊息。

CarouselTemplate 輪播樣板

ConfirmTemplate 確認樣板會提供最多十個群組 (按鈕樣板) 供使用者左右滑動點選，相關屬性如下：

屬性	說明
columns	要出現的按鈕樣板，使用串列格式。
thumbnailImageUrl	縮圖連結，支援 jpg 和 png，最大寬度 1024px。
imageAspectRatio	圖片比例，預設 rectangle (1.51:1)，可設定 square (1:1)。
imageSize	圖片尺寸，預設 cover (絕對撐滿，才切超過顯示比例的區域)，可設定 contain。
imageBackgroundColor	放置圖片區域的背景顏色，預設白色 #FFFFFF。
title	樣板標題。
text	樣板說明文字。
actions	點擊按鈕觸發的行為，一個按鈕一種行為，最多支援四個按鈕。

import TemplateSendMessage 和 CarouselTemplate 函式庫之後，就能透過 CarouselTemplate() 方法發送輪播樣板訊息，與按鈕樣板不同的是，**輪播樣板會先使用 columns 放置要輪播的按鈕樣板，每個按鈕樣板放在 CarouselColumn 裡，而按鈕同樣是透過 actions 屬性設定，actions 屬性為 list 格式，長度最多為 2，內容是 MessageAction (文字)。**

下方的程式碼範例中，使用 line_bot_api.push_message() 方法，搭配 CarouselTemplate() 方法發送確認樣板。

```
from linebot import LineBotApi, WebhookHandler
# 需要額外載入對應的函示庫
from linebot.models import MessageAction, TemplateSendMessage,
CarouselTemplate,  CarouselColumn
line_bot_api = LineBotApi('你的 Channel access token')
line_bot_api.push_message('你的 user ID', TemplateSendMessage(
```

```
    alt_text='CarouselTemplate',
    template=CarouselTemplate(
        columns=[
            CarouselColumn(

thumbnail_image_url='https://steam.oxxostudio.tw/download/python/line-
template-message-demo.jpg',
                title=' 選單 1',
                text=' 說明文字 1',
                actions=[
                    PostbackAction(
                        label='postback',
                        data='data1'
                    ),
                    MessageAction(
                        label='hello',
                        text='hello'
                    ),
                    URIAction(
                        label='oxxo.studio',
                        uri='http://oxxo.studio'
                    )
                ]
            ),
            CarouselColumn(

thumbnail_image_url='https://steam.oxxostudio.tw/download/python/line-
template-message-demo2.jpg',
                title=' 選單 2',
                text=' 說明文字 2',
                actions=[
                    PostbackAction(
                        label='postback',
                        data='data1'
                    ),
                    MessageAction(
                        label='hi',
                        text='hi'
                    ),
```

```
                   URIAction(
                       label='STEAM 教育學習網 ',
                       uri='https://steam.oxxostudio.tw'
                   )
               ]
           )
       ]
   )
))
```

（範例程式碼：ch6/code03.py）

程式執行後，就會看見 LINE BOT 向自己發送內含兩組按鈕樣板的輪播樣板，可以用手指左右滑動切換。

ImageCarouselTemplate 圖片輪播樣板

ImageCarouselTemplate 圖片輪播樣板會提供「最多十張」可以滑動選擇的圖片，作為按鈕供使用者點選，相關屬性如下：

屬性	說明
text	樣板說明文字。
columns	要出現的圖片樣板，使用串列格式。
imageUrl	圖片網址，支援 jpg 和 png，最大寬度 1024px。
actions	點擊圖片所觸發的行為，一張圖片一種行為，最多支援十張圖片。

import TemplateSendMessage 和 ImageCarouselTemplate 函式庫之後，就能透過 ImageCarouselTemplate() 方法發送圖片樣板訊息，**圖片樣板會先使用 columns 放置要輪播的圖片，每張圖片放在 ImageCarouselColumn 裡，而點擊圖片的說明文字是透過 actions 屬性設定。**

下方的程式碼範例中，使用 line_bot_api.push_message() 方法，搭配 ImageCarouselTemplate() 方法發送確認樣板。

```
from linebot import LineBotApi, WebhookHandler
# 需要額外載入對應的函示庫
from linebot.models import MessageAction, TemplateSendMessage,
ImageCarouselTemplate, ImageCarouselColumn
line_bot_api = LineBotApi('你的 Channel access token')
line_bot_api.push_message('你的 user ID', TemplateSendMessage(
    alt_text='ImageCarousel template',
    template=ImageCarouselTemplate(
        columns=[
            ImageCarouselColumn(
                image_url='https://upload.wikimedia.org/wikipedia/en/
a/a6/Pok%C3%A9mon_Pikachu_art.png',
                action=MessageAction(
```

```
                    label=' 皮卡丘 ',
                    text=' 皮卡丘 '
                )
            ),
        ImageCarouselColumn(
            image_url='https://upload.wikimedia.org/wikipedia/
en/5/59/Pok%C3%A9mon_Squirtle_art.png',
            action=MessageAction(
                label=' 傑尼龜 ',
                text=' 傑尼龜 '
            )
        )
    ]
    )
))
```

(範例程式碼：ch6/code04.py)

程式執行後，就會看見 LINE BOT 向自己發送內含兩張圖片的圖片輪播樣板，可以用手指左右滑動切換 (圖片來源為維基百科)。

6-2　發送 Flex Message

LINE BOT 可以使用 Flex Message (彈性樣板訊息) 發送客製化的選單訊息，Flex Message 使用網頁 CSS3 的 Flex 語法，可以設計較豐富且多樣性的按鈕版型，此外，透過 Flex Message 設計工具，不僅能直接修改範本，也能輕鬆的從無到有設計複雜的版面，讓 LINE BOT 回應的訊息更具特色，接下來會介紹如何發送 Flex Message 以及設計客製化的選單。

設計 Flex Message

開啟 LINE 官方提供的設計工具：Flex Message Simulator，並使用 LINE 的帳號登入，登入後預設會開啟一個展示的樣板。

> Flex Message Simulator：https://developers.line.biz/console/fx/

設計工具主要分成三個區塊：

- 左邊的區塊：預覽畫面 (傳送到 LINE 的訊息長相)。

- 中間的區塊：Flex Message 的樹狀清單結構。

- 右邊的區塊：點擊中間區塊清單裡的元件，出現對應可以修改的屬性或參數。

　　預覽畫面　　　　　　　　區塊樹狀清單　　　　　　參數與屬性修改

點擊中間清單的元件後，就能從上方的「+」號，加入額外的元件。

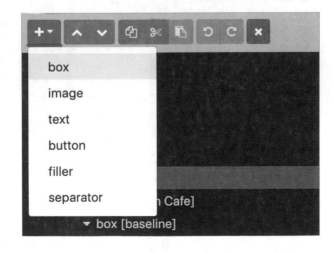

右上方有四個功能按鈕：

● New：建立新的 Flex Message 樣板。

● Showcase：使用範本。

● Send：將設計好的樣板傳送到 LINE 測試。

● View as JSON：查看發送樣板的 JSON 檔案。

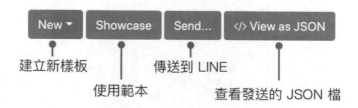

點擊右上方的 Send 按鈕後，會出現 Register destination 的按鈕，繼續點擊後會出現 QRCode，掃描 QRCode 加入好友就能測試。

Python 發送 Flex Message

點擊 Flex Message Simulator 右上方的 View as JSON 按鈕,開啟 Flex Message 的 JSON 檔案。

```json
{
 "type": "bubble",
 "hero": {
  "type": "image",
  "url": "https://scdn.line-apps.com/n/channel_devcenter/img/fx/01_1_cafe.png",
  "size": "full",
  "aspectRatio": "20:13",
  "aspectMode": "cover",
  "action": {
   "type": "uri",
   "uri": "http://linecorp.com/"
  }
 },
 "body": {
  "type": "box",
  "layout": "vertical",
  "contents": [
   {
    "type": "text",
    "text": "Brown Cafe",
    "weight": "bold",
    "size": "xl"
   },
   {
    "type": "box",
    "layout": "baseline",
    "margin": "md",
    "contents": [
     {
      "type": "icon",
      "size": "sm",
      "url": "https://scdn.line-apps.com/n/channel_devcenter/img/fx/review_gold_star_28.png"
     },
```

JSON spec

Copy　Close　Apply

import FlexSendMessage、BubbleContaine 和 ImageComponent 方法,填入自己的 access token 和 user ID,並將剛剛產生的 Flex Message JSON 內容貼入 FlexSendMessage 方法的 contents 屬性。

```python
from linebot import LineBotApi, WebhookHandler
# 載入對應的函式庫
from linebot.models import FlexSendMessage, BubbleContainer,
ImageComponent
line_bot_api = LineBotApi('你的 Access Token')
# 剛剛 Flex Message 的 JSON 檔案就貼在下方
line_bot_api.push_message('你的 User ID', FlexSendMessage(
    alt_text='hello',
    contents={
      "type": "bubble",
      "hero": {
        "type": "image",
        "url": "https://scdn.line-apps.com/n/channel_devcenter/img/
fx/01_1_cafe.png",
        "size": "full",
        "aspectRatio": "20:13",
        "aspectMode": "cover",
        "action": {
          "type": "uri",
          "uri": "http://linecorp.com/"
        }
      },
      "body": {
        "type": "box",
        "layout": "vertical",
        "contents": [
          {
            "type": "text",
            "text": "Brown Cafe",
            "weight": "bold",
            "size": "xl"
          },
          {
            "type": "box",
            "layout": "baseline",
            "margin": "md",
            "contents": [
              {
                "type": "icon",
```

```json
          "size": "sm",
          "url": "https://scdn.line-apps.com/n/channel_devcenter/img/fx/review_gold_star_28.png"
        },
        {
          "type": "icon",
          "size": "sm",
          "url": "https://scdn.line-apps.com/n/channel_devcenter/img/fx/review_gold_star_28.png"
        },
        {
          "type": "icon",
          "size": "sm",
          "url": "https://scdn.line-apps.com/n/channel_devcenter/img/fx/review_gold_star_28.png"
        },
        {
          "type": "icon",
          "size": "sm",
          "url": "https://scdn.line-apps.com/n/channel_devcenter/img/fx/review_gold_star_28.png"
        },
        {
          "type": "icon",
          "size": "sm",
          "url": "https://scdn.line-apps.com/n/channel_devcenter/img/fx/review_gray_star_28.png"
        },
        {
          "type": "text",
          "text": "4.0",
          "size": "sm",
          "color": "#999999",
          "margin": "md",
          "flex": 0
        }
      ]
    },
    {
```

```
"type": "box",
"layout": "vertical",
"margin": "lg",
"spacing": "sm",
"contents": [
  {
    "type": "box",
    "layout": "baseline",
    "spacing": "sm",
    "contents": [
      {
        "type": "text",
        "text": "Place",
        "color": "#aaaaaa",
        "size": "sm",
        "flex": 1
      },
      {
        "type": "text",
        "text": "Miraina Tower, 4-1-6 Shinjuku, Tokyo",
        "wrap": True,
        "color": "#666666",
        "size": "sm",
        "flex": 5
      }
    ]
  },
  {
    "type": "box",
    "layout": "baseline",
    "spacing": "sm",
    "contents": [
      {
        "type": "text",
        "text": "Time",
        "color": "#aaaaaa",
        "size": "sm",
        "flex": 1
      },
```

```
                {
                    "type": "text",
                    "text": "10:00 - 23:00",
                    "wrap": True,
                    "color": "#666666",
                    "size": "sm",
                    "flex": 5
                }
            ]
        }
    ]
    }
    ]
},
"footer": {
    "type": "box",
    "layout": "vertical",
    "spacing": "sm",
    "contents": [
        {
            "type": "button",
            "style": "link",
            "height": "sm",
            "action": {
                "type": "uri",
                "label": "CALL",
                "uri": "https://linecorp.com"
            }
        },
        {
            "type": "button",
            "style": "link",
            "height": "sm",
            "action": {
                "type": "uri",
                "label": "WEBSITE",
                "uri": "https://linecorp.com"
            }
        },
```

```
        {
          "type": "box",
          "layout": "vertical",
          "contents": [],
          "margin": "sm"
        }
      ],
      "flex": 0
    }
  }
))
```
(範例程式碼：ch6/code05.py)

程式執行後，就會看見 LINE BOT 發送出 Flex Message。

6-3　建立圖文選單

在行動版的 LINE 裡，有時會出現「圖文選單」的功能 (最下方可以點選的圖文區塊)，圖文選單為行動裝置的操作者，提供了相當方便的使用體驗，因為使用點擊發送訊息的速度，遠比用手指打字來得迅速，活用圖文選單，就能替 LINE BOT 增加更多好用的功能。接下來將會介紹如何在 LINE 官方帳號開啟基本圖文選單功能，以及如何透過 Python 程式產生客製化的圖文選單。

什麼是 LINE 圖文選單？

INE 圖文選單位於行動版的聊天室畫面下方，是一個可以彈出與收合的選單，使用者不管是因為推播或其他互動打開您的官方帳號時，都能看到這個選單，舉凡「外導連結」、「優惠券」、「集點卡」及「關鍵字」都能設定。

以下圖為例，下方台新知識王的區塊，就是圖文選單。

LINE 官方帳號建立圖文選單

進入 LINE 的官方帳號，選擇要建立圖文選單的官方帳號 (官方帳號可能是手動建立，或建立 LINE BOT 時自動產生)。

左側選單選擇「圖文選單」，點擊「建立圖文選單」。

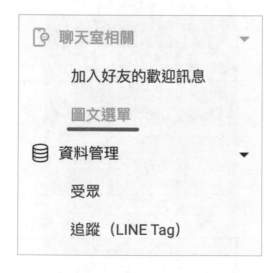

建立時需要填寫「標題」、「時間」、「選單顯示的文字」和「預設顯示方式」。

填寫完畢後，進入內容設定，點擊「選擇版型」。

內容設定

請選擇版型並上傳背景圖片。

選擇版型

上傳圖片

建立圖片

從預設的版型中，挑選需要的樣式，樣式裡每個格子都是一個「按鈕」，也就是使用者可以點擊的區域。

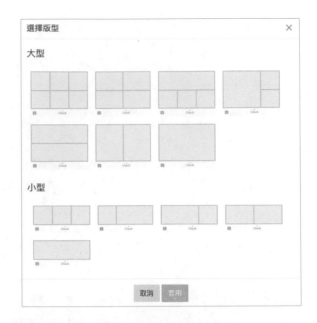

樣式設定後,就能替樣式裡每個按鈕格子加入圖片 (或使用單一張背景圖)。

最後在「動作」的位置,設定點擊每個按鈕時要進行的動作,可以設定發送
文字、開啟連結 ... 等功能。

完成後按下建立，在設定的時間範圍裡從 LINE 開啟與官方帳號的聊天視窗，就會看見下方出現圖文選單（有時需要等待一兩分鐘才會出現），點擊選單就會出現設定的動作。

👆 Python 建立圖文選單（準備圖片）

除了透過 LINE 官方帳號建立圖文選單，也可以按照下列步驟，使用 Python 在本機環境設計出更有創意的圖文選單樣式，首先準備一張背景圖片，圖片的格式要求如下：

● 寬度：800 ～ 2500 px(實際大小取決於圖文選單設定)

● 高度：250 ～ 1686 px (實際大小取決於圖文選單設定)

● 長寬比：保持 1.45 ～ 1.5

● 格式：jpg 或 png

範例圖片為一個商場的插畫，裡面包含七個按鈕，記錄每個按鈕的 xy 座標以及長寬，準備好圖片後，將圖片上傳到 Google Drive 雲端空間或下載到電腦資料夾，圖片位置和 Colab 專案或 python 檔案放在一起。

範例圖片網址：

https://steam.oxxostudio.tw/download/python/line-rich-menu-demo.jpg

---

Python 建立圖文選單 (程式設定)

使用本機環境或 Colab，建立一個專案，輸入下方指令安裝 line-bot-sdk (
Anaconda Jupyter 和 Colab 需要使用 !pip)。

```
!pip install line-bot-sdk
```

按照 LINE 官方提供的方法，使用下方的程式碼 (已經按照範例圖片設定好
各個按鈕的 areas 大小和位置)，只要輸入自己的 Channel access token 和
Channel secret，執行後就會使用 POST 的方法向指定網址發送 request，建
立圖文選單。

```
import requests
import json
# 設定 headers，輸入你的 Access Token，記得前方要加上「Bearer 」( 有一個空白 )
headers = {'Authorization':'Bearer 你的 Access Token','Content-
Type':'application/json'}

body = {
    'size': {'width': 2500, 'height': 1686},     # 設定尺寸
    'selected': 'true',                          # 預設是否顯示
    'name': 'Richmenu demo',                     # 選單名稱
    'chatBarText': 'Richmenu demo',              # 選單在 LINE 顯示的標題
    'areas':[                                    # 選單內容
        {
            'bounds': {'x': 341, 'y': 75, 'width': 560, 'height': 340},
# 選單位置與大小
            'action': {'type': 'message', 'text': '電器'} # 點擊後傳送文字
        },
        {
            'bounds': {'x': 1434, 'y': 229, 'width': 930, 'height': 340},
            'action': {'type': 'message', 'text': '運動用品'}
        },
        {
            'bounds': {'x': 122, 'y': 641, 'width':560, 'height': 340},
            'action': {'type': 'message', 'text': '客服'}
        },
```

```
        {
          'bounds': {'x': 1012, 'y': 645, 'width': 560, 'height': 340},
          'action': {'type': 'message', 'text': '餐廳'}
        },
        {
          'bounds': {'x': 1813, 'y': 677, 'width': 560, 'height': 340},
          'action': {'type': 'message', 'text': '鞋子'}
        },
        {
          'bounds': {'x': 423, 'y': 1203, 'width': 560, 'height': 340},
          'action': {'type': 'message', 'text': '美食'}
        },
        {
          'bounds': {'x': 1581, 'y': 1133, 'width': 560, 'height': 340},
          'action': {'type': 'message', 'text': '衣服'}
        }
    ]
  }
# 向指定網址發送 request
req = requests.request('POST', 'https://api.line.me/v2/bot/richmenu',
                       headers=headers,data=json.dumps(body).
encode('utf-8'))
# 印出得到的結果
print(req.text)
```
(範例程式碼：ch6/code06.py)

執行程式後就會建立圖文選單，並得到一個圖文選單的 ID (Rich menu ID)。

```
{"richMenuId":"richmenu-████████████████████████████"}
```

取得圖文選單的 ID 後，編輯新的一組程式碼（Colab 或 Jupyter 可以新增另外一個程式碼編輯區塊，如果是 Colab 需要轉換路徑），輸入 Channel Access Token 和剛剛產生的 Rich menu ID，執行後就會將圖片上傳到 LINE 並與剛剛的圖文選單綁定。

```python
from linebot import LineBotApi, WebhookHandler

line_bot_api = LineBotApi('你的 Access Token')

# import os
# os.chdir('/content/drive/MyDrive/Colab Notebooks')  # Colab 換路徑使用

# 開啟對應的圖片
with open('demo.jpg', 'rb') as f:
    line_bot_api.set_rich_menu_image('你的圖文選單 ID', 'image/jpeg', f)
```
（範例程式碼：ch6/code07.py）

最後，再編輯新的一組程式碼（Colab 或 Jupyter 可以新增另外一個程式碼編輯區塊），輸入 Channel Access Token 和剛剛產生的 Rich menu ID，完成並執行後，在行動裝置上開啟 LINE BOT 的聊天視窗，下方就會出現圖文選單，點擊圖文選單，就會發送指定的文字訊息。

```python
import requests

headers = {'Authorization':'Bearer 你的 Access Token'}

req = requests.request('POST', 'https://api.line.me/v2/bot/user/all/
richmenu/ 你的圖文選單 ID', headers=headers)

print(req.text)
```
（範例程式碼：ch6/code08.py）

Python 其他圖文選單方法

除了建立圖片選單，還有其他的方法可以操作圖文選單，下方的程式碼執行後，可以取得目前所有已經建立的圖文選單 ID。

參考：

https://developers.line.biz/en/reference/messaging-api/#rich-menu

---

```
from linebot import  LineBotApi, WebhookHandler
line_bot_api = LineBotApi('你的 Access Token')
rich_menu_list = line_bot_api.get_rich_menu_list()
for rich_menu in rich_menu_list:
    print(rich_menu.rich_menu_id)
```

下方的程式碼可以根據圖文選單的 ID，刪除指定的圖文選單。

```
from linebot import LineBotApi, WebhookHandler
line_bot_api = LineBotApi('你的 Access Token')
line_bot_api.delete_rich_menu('圖文選單 ID')
```

圖文選單物件格式

之前介紹的程式碼只使用了「message」的格式進行發送，但 LINE 其實提供了許多圖文選單的物件格式，例如 uri（超連結）、location（地點）... 等，舉例來說，如果將上方的範例中「運動用品」的內容改成下方的程式碼，執行後，點擊運動用品按鈕時，就會開啟網頁。

> 參考：
>
> https://developers.line.biz/en/reference/messaging-api/#rich-menu

```
{
  'bounds': {'x': 1434, 'y': 229, 'width': 930, 'height': 340},
  'action': {'type': 'uri', 'label': '運動用品', 'uri':'https://www.
oxxostudio.tw'} # 點擊後開啟網頁
},
```

6-4 切換圖文選單

由於 LINE 可以設定「多組」圖文選單，每一組圖文選單之間都可透過指令切換（有點類似切換網頁選單的概念），接下來會介紹如何設定多組圖文選單、設定圖文選單別名 Alias ID，以及切換不同的圖文選單。

設定圖文選單 A

下載圖文選單 A 的範例圖片，將圖片放在與程式同樣的資料夾裡（Colab 放在與程式相同的雲端資料夾裡）。

下載 A 圖片：

https://steam.oxxostudio.tw/download/python/line-rich-menu-switch-demo-a.jpg

使用下方的程式，輸入自己的 access token，設定第一個圖文選單（選單 A）的內容，設定的重點如下：

● name 設定為 aaa，表示別名 Alias Id。

- 預設有三個按鈕，第一個連結到選單 A 按鈕行為設定為 postback，點擊後會在背後送出 postback，不會有顯示的反應。

- 連結到選單 B 和 C 的按鈕，設定為 richmenuswitch，richMenuAliasId 設定為 bbb 和 ccc。

```python
import requests, json
headers = {'Authorization':'Bearer 你的 access token','Content-
Type':'application/json'}

body = {
    'size': {'width': 2500, 'height': 1200},# 設定尺寸
    'selected': 'true',                     # 預設是否顯示
    'name': 'aaa',                          # 選單名稱（別名 Alias Id）
    'chatBarText': ' 選單 A',               # 選單在 LINE 顯示的標題
    'areas':[                               # 選單內容
        {
          'bounds': {'x': 0, 'y': 0, 'width': 830, 'height': 280},
          'action': {'type': 'postback', 'data':'no-data'}  # 按鈕 A 使
用 postback
        },
        {
          'bounds': {'x': 835, 'y': 0, 'width':830, 'height': 640},
          'action': {'type': 'richmenuswitch', 'richMenuAliasId':
'bbb', 'data':'change-to-bbb'} # 按鈕 B 使用 richmenuswitch
        },
        {
          'bounds': {'x': 1670, 'y': 0, 'width':830, 'height': 640},
          'action': {'type': 'richmenuswitch', 'richMenuAliasId':
'ccc', 'data':'change-to-ccc'} # 按鈕 C 使用 richmenuswitch
        }
    ]
  }
req = requests.request('POST', 'https://api.line.me/v2/bot/richmenu',
                  headers=headers,data=json.dumps(body).
encode('utf-8'))
print(req.text)
```

（範例程式碼：ch6/code09.py）

上方程式碼執行後，會得到圖文選單 A 的 ID，取得圖文選單的 ID 後，接著編輯新的一組程式碼（Colab 或 Jupyter 可以新增另外一個程式碼編輯區塊，如果是 Colab 需要轉換路徑），輸入 Channel Access Token 和剛剛產生的 Rich menu ID，執行後就會將圖片上傳到 LINE 並與剛剛的圖文選單綁定。

```
from linebot import LineBotApi, WebhookHandler

line_bot_api = LineBotApi(' 你的 Access Token')

# import os
# os.chdir('/content/drive/MyDrive/Colab Notebooks')  # Colab 換路徑使用

# 開啟對應的圖片
with open('line-rich-menu-switch-demo-a.jpg', 'rb') as f:
    line_bot_api.set_rich_menu_image(' 你的圖文選單 ID', 'image/jpeg', f)
```
（ 範例程式碼：ch6/code10.py ）

執行下方程式碼，編輯新的一組程式碼（ Colab 或 Jupyter 可以新增另外一個程式碼編輯區塊 ），輸入 Channel Access Token 和剛剛產生的 Rich menu ID 進行綁定。

```
import requests
import json
headers = {'Authorization':'Bearer 你的 access token','Content-
Type':'application/json'}
body = {
    "richMenuAliasId":"aaa",
    "richMenuId":" 圖文選單 id"
}
req = requests.request('POST', 'https://api.line.me/v2/bot/richmenu/alias',
                    headers=headers,data=json.dumps(body).
encode('utf-8'))
print(req.text)
```
（ 範例程式碼：ch6/code11.py ）

最後執行下方程式碼，將圖文選單傳送到對應的 LINE 機器人。

```
import requests
headers = {"Authorization":"Bearer 你的 access token","Content-
Type":"application/json"}
req = requests.request('POST', 'https://api.line.me/v2/bot/user/all/
richmenu/ 圖文選單 id', headers=headers)
print(req.text)
```
(範例程式碼：ch6/code12.py)

設定圖文選單 B

下載圖文選單 B 的範例圖片，將圖片放在與程式同樣的資料夾裡 (Colab 放在與程式相同的雲端資料夾裡)。

下載 B 圖片：

https://steam.oxxostudio.tw/download/python/line-rich-menu-switch-demo-b.jpg

使用下方的程式，輸入自己的 access token，設定第一個圖文選單 (選單 B) 的內容，設定的重點如下：

● name 設定為 bbb，表示別名 Alias Id。

● 預設有三個按鈕，第一個連結到選單 A 按鈕行為設定為 postback，點擊後會在背後送出 postback，不會有顯示的反應。

● 連結到選單 A 和 C 的按鈕，設定為 richmenuswitch，richMenuAliasId 設
  定為 aaa 和 ccc。

```python
import requests, json
headers = {'Authorization':'Bearer 你的 access token','Content-
Type':'application/json'}

body = {
    'size': {'width': 2500, 'height': 1200},# 設定尺寸
    'selected': 'true',                      # 預設是否顯示
    'name': 'bbb',                           # 選單名稱 ( 別名 Alias Id )
    'chatBarText': ' 選單 B',                # 選單在 LINE 顯示的標題
    'areas':[                                # 選單內容
        {
           'bounds': {'x': 0, 'y': 0, 'width': 830, 'height': 280},
           'action': {'type': 'richmenuswitch', 'richMenuAliasId':
'aaa', 'data':'change-to-aaa'} # 按鈕 A 使用 richmenuswitch
        },
        {
           'bounds': {'x': 835, 'y': 0, 'width':830, 'height': 640},
           'action': {'type': 'postback', 'data':'no-data'}    # 按鈕 B
使用 postback
        },
        {
           'bounds': {'x': 1670, 'y': 0, 'width':830, 'height': 640},
           'action': {'type': 'richmenuswitch', 'richMenuAliasId':
'ccc', 'data':'change-to-ccc'} # 按鈕 C 使用 richmenuswitch
        }
    ]
  }
req = requests.request('POST', 'https://api.line.me/v2/bot/richmenu',
                   headers=headers,data=json.dumps(body).
encode('utf-8'))
print(req.text)
```
(範例程式碼：ch6/code13.py)

上方程式碼執行後，會得到圖文選單 B 的 ID，取得圖文選單的 ID 後，接著
編輯新的一組程式碼（Colab 或 Jupyter 可以新增另外一個程式碼編輯區塊，
如果是 Colab 需要轉換路徑），輸入 Channel Access Token 和剛剛產生的
Rich menu ID，執行後就會將圖片上傳到 LINE 並與剛剛的圖文選單綁定。

```python
from linebot import LineBotApi, WebhookHandler

line_bot_api = LineBotApi('你的 Access Token')

# import os
# os.chdir('/content/drive/MyDrive/Colab Notebooks') # Colab 換路徑使用

# 開啟對應的圖片
with open('line-rich-menu-switch-demo-b.jpg', 'rb') as f:
    line_bot_api.set_rich_menu_image('你的圖文選單 ID', 'image/jpeg', f)
```
（範例程式碼：ch6/code14.py）

執行下方程式碼，編輯新的一組程式碼（Colab 或 Jupyter 可以新增另外一
個程式碼編輯區塊），輸入 Channel Access Token 和剛剛產生的 Rich menu
ID 進行綁定。

```python
import requests
import json
headers = {'Authorization':'Bearer 你的 access token','Content-
Type':'application/json'}
body = {
    "richMenuAliasId":"bbb",
    "richMenuId":"圖文選單 id"
}
req = requests.request('POST', 'https://api.line.me/v2/bot/richmenu/alias',
                    headers=headers,data=json.dumps(body).
encode('utf-8'))
print(req.text)
```
（範例程式碼：ch6/code15.py）

6 傳送不同類型的 LINE 訊息

最後執行下方程式碼，將圖文選單傳送到對應的 LINE 機器人。

```
import requests
headers = {"Authorization":"Bearer 你的 access token","Content-
Type":"application/json"}
req = requests.request('POST', 'https://api.line.me/v2/bot/user/all/
richmenu/圖文選單 id', headers=headers)
print(req.text)
```
（範例程式碼：ch6/code12.py）

設定圖文選單 C

下載圖文選單 C 的範例圖片，將圖片放在與程式同樣的資料夾裡（Colab 放在與程式相同的雲端資料夾裡）。

下載 C 圖片：
https://steam.oxxostudio.tw/download/python/line-rich-menu-switch-demo-c.jpg

使用下方的程式，輸入自己的 access token，設定第一個圖文選單（選單 C）的內容，設定的重點如下：

● name 設定為 ccc，表示別名 Alias Id。

6-38

- 預設有三個按鈕，第一個連結到選單 A 按鈕行為設定為 postback，點擊後會在背後送出 postback，不會有顯示的反應。

- 連結到選單 A 和 B 的按鈕，設定為 richmenuswitch，richMenuAliasId 設定為 aaa 和 bbb。

```python
import requests, json
headers = {'Authorization':'Bearer 你的 access token','Content-
Type':'application/json'}

body = {
    'size': {'width': 2500, 'height': 1200},# 設定尺寸
    'selected': 'true',                     # 預設是否顯示
    'name': 'ccc',                          # 選單名稱 ( 別名 Alias Id )
    'chatBarText': ' 選單 C',                # 選單在 LINE 顯示的標題
    'areas':[                               # 選單內容
        {
          'bounds': {'x': 0, 'y': 0, 'width': 830, 'height': 280},
          'action': {'type': 'richmenuswitch', 'richMenuAliasId':
'aaa', 'data':'change-to-aaa'} # 按鈕 A 使用 richmenuswitch
        },
        {
          'bounds': {'x': 835, 'y': 0, 'width':830, 'height': 640},
          'action': {'type': 'richmenuswitch', 'richMenuAliasId':
'bbb', 'data':'change-to-ccc'} # 按鈕 B 使用 richmenuswitch
        },
        {
          'bounds': {'x': 1670, 'y': 0, 'width':830, 'height': 640},
          'action': {'type': 'postback', 'data':'no-data'}  # 按鈕 C 使
用 postback
        }
    ]
  }
req = requests.request('POST', 'https://api.line.me/v2/bot/richmenu',
                  headers=headers,data=json.dumps(body).
encode('utf-8'))
print(req.text)
```
(範例程式碼：ch6/code16.py)

上方程式碼執行後，會得到圖文選單 C 的 ID，取得圖文選單的 ID 後，接著編輯新的一組程式碼 (Colab 或 Jupyter 可以新增另外一個程式碼編輯區塊，如果是 Colab 需要轉換路徑)，輸入 Channel Access Token 和剛剛產生的 Rich menu ID，執行後就會將圖片上傳到 LINE 並與剛剛的圖文選單綁定。

```python
from linebot import LineBotApi, WebhookHandler

line_bot_api = LineBotApi('你的 Access Token')

# import os
# os.chdir('/content/drive/MyDrive/Colab Notebooks') # Colab 換路徑使用

# 開啟對應的圖片
with open('line-rich-menu-switch-demo-c.jpg', 'rb') as f:
    line_bot_api.set_rich_menu_image('你的圖文選單 ID', 'image/jpeg', f)
```
(範例程式碼：ch6/code17.py)

執行下方程式碼，編輯新的一組程式碼 (Colab 或 Jupyter 可以新增另外一個程式碼編輯區塊)，輸入 Channel Access Token 和剛剛產生的 Rich menu ID 進行綁定。

```python
import requests
import json
headers = {'Authorization':'Bearer 你的 access token','Content-
Type':'application/json'}
body = {
    "richMenuAliasId":"ccc",
    "richMenuId":"圖文選單 id"
}
req = requests.request('POST', 'https://api.line.me/v2/bot/richmenu/alias',
                       headers=headers,data=json.dumps(body).
encode('utf-8'))
print(req.text)
```
(範例程式碼：ch6/code18.py)

最後執行下方程式碼，將圖文選單傳送到對應的 LINE 機器人。

```
import requests
headers = {"Authorization":"Bearer 你的 access token","Content-
Type":"application/json"}
req = requests.request('POST', 'https://api.line.me/v2/bot/user/all/
richmenu/圖文選單 id', headers=headers)
print(req.text)
```
（範例程式碼：ch6/code12.py）

測試執行結果

三個選單都完成後，開啟 LINE 機器人，在對話視窗的下方就會出現圖文選
單，點擊按鈕就能切換圖文選單。

效果預覽：

https://steam.oxxostudio.tw/webp/python/example/line-rich-menu-switch-04.gif

 小結

LINE 在行動裝置上的功能，遠比電腦版本來得多元，只要熟悉了樣板訊息、Flex Message 和圖文選單，就能提供更多方便的使用體驗，因為使用點擊發送訊息的速度，遠比用手指打字來得迅速，活用這些功能，就能大幅增加 LINE BOT 的實用度。

Note

7

實作 LINE
氣象機器人

了解 LINE 的訊息傳遞方式與各種格式之後，接著就要開始實作 LINE 機器人，接下來的這個章節，會從簡單的氣象爬蟲開始介紹，一步步實作出一個可以詢問氣象資訊的「氣象機器人」。

本章節的範例程式碼：

https://github.com/oxxostudio/book-code/tree/master/linebot/ch7

兩點注意事項：

要實作 LINE 的氣象機器人需要先註冊 LINE 開發者帳號（取得 Access Token 和 Channel secret），以及架設 Webhook 與 LINE 串接，請先閱讀本書第二章「建立 LINE BOT」和第三章「開發環境設定＆串接 LINE BOT」，完成相關步驟後再進行閱讀。

實作過程中會需要使用一些 Python 基本語法，例如函式、串列與字典解析、字串格式化、邏輯、迴圈 ... 等，如果對於語法不熟悉，請前往 https://steam.oxxostudio.tw/category/python/ 進行相關基礎語法學習。

7-1　氣象機器人 (1) - 雷達回波與地震資訊

接下來的這個章節，會先使用簡單的網路爬蟲技巧，取得雷達回波圖與地震資訊，再將相關資訊，透過 LINE BOT 聊天的方式，傳遞給使用者。

👆 建立測試用的 Webhook

執行下方的程式碼產生 Webhook，回到 LINE Channel，確認 Webhook 可以正常運作 (驗證 Verify 後出現 success 表示 Webhook 沒有問題)。

> 注意 ! 如果使用 ngrok 搭配 Colab 做開發測試，每次執行後產生的網址都不同，
> 需要重複更新 LINE 的 Webhook。

```python
# Colab 使用，本機環境請刪除
from flask_ngrok import run_with_ngrok

from flask import Flask, request

# 載入 LINE Message API 相關函式庫
from linebot import LineBotApi, WebhookHandler
from linebot.models import MessageEvent, TextMessage, TextSendMessage
# 載入 json 標準函式庫，處理回傳的資料格式
import json

app = Flask(__name__)

@app.route("/", methods=['POST'])
def linebot():
    # 取得收到的訊息內容
    body = request.get_data(as_text=True)
    try:
        line_bot_api = LineBotApi('你的 LINE Channel access token')
        handler = WebhookHandler('你的 LINE Channel secret')
```

```
        signature = request.headers['X-Line-Signature']
        handler.handle(body, signature)
        # 轉換內容為 json 格式
        json_data = json.loads(body)
        print(json_data)
    except:
        print('error')
    return 'OK'
if __name__ == "__main__":
    # Colab 使用，本機環境請刪除
    run_with_ngrok(app)
    app.run()
```

(範例程式碼：ch7/code01.py)

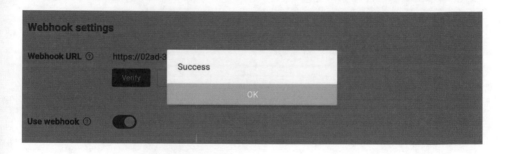

回傳雷達回波圖

雷達回波是由雷達發射之電磁波經由大氣中的降水粒子 (雨、雪、冰雹 ... 等) 反射回來的訊號，根據雷達接收到降水粒子所反射回來的訊號強度，再利用不同顏色顯示，就成為了雷達回波圖，雷達回波圖除了可以直接從中央氣象局的網站取得，在政府資料開放平臺也有提供即時的圖片資料。

- 關於雷達回波圖：https://data.gov.tw/dataset/75125
- 臺灣 (鄰近區域)_ 無地形：
  https://cwbopendata.s3.ap-northeast-1.amazonaws.com/MSC/O-A0058-001.png
- 臺灣 (較大範圍)_ 無地形：
  https://cwbopendata.s3.ap-northeast-1.amazonaws.com/MSC/O-A0058-003.png

由於雷達回波圖的圖片網址是固定的，所以只要修改原本的程式碼，判斷
LINE BOT 收到的訊息為「雷達回波圖」或「雷達回波」時，就使用 reply
message 的方法回傳圖片訊息，程式碼的重點如下 (詳細說明寫在程式碼的
註解中)：

- 建立了一個 reply_image 函式，使用 requests 的 GET 方法傳送圖片。
- 因為傳送訊息時需要 access token 和 reply token，所以將這兩個值變成
  reply_message 函式的參數。
- 將 access token 和 channel_secret 獨立為變數。
- 使用 if 判斷訊息是否出現「雷達回波圖」或「雷達回波」。

實作過程中，可能會遇到如果傳送過圖片，過一陣子再次傳送時，顯示的是
「舊」的雷達回波圖，而非「即時」雷達回波圖，這是因為 LINE 認為「同一個
網址」就應該是「同一張圖片」(API 裡的圖片網址不會變，只有圖片內容變了)，
所以在傳送時可以使用 time.time_ns() 方法，替圖片加上「時間參數」，就能避
開這個問題。

```
# Colab 使用，本機環境請刪除
from flask_ngrok import run_with_ngrok

from flask import Flask, request
```

```python
# 載入 LINE Message API 相關函式庫
from linebot import LineBotApi, WebhookHandler
from linebot.models import MessageEvent, TextMessage, TextSendMessage

# 載入 json 標準函式庫，處理回傳的資料格式
import requests, json, time

app = Flask(__name__)

access_token = '你的 LINE Channel access token'
channel_secret = '你的 LINE Channel secret'

@app.route("/", methods=['POST'])
def linebot():
    # 取得收到的訊息內容
    body = request.get_data(as_text=True)
    try:
        line_bot_api = LineBotApi(access_token)
        handler = WebhookHandler(channel_secret)
        signature = request.headers['X-Line-Signature']
        handler.handle(body, signature)
        # 轉換內容為 json 格式
        json_data = json.loads(body)
        # 取得回傳訊息的 Token ( reply message 使用 )
        reply_token = json_data['events'][0]['replyToken']
        # 取得使用者 ID ( push message 使用 )
        user_id = json_data['events'][0]['source']['userId']
        print(json_data)
        # 如果傳送的是 message
        if 'message' in json_data['events'][0]:
            # 如果 message 的類型是文字 text
            if json_data['events'][0]['message']['type'] == 'text':
                # 取出文字
                text = json_data['events'][0]['message']['text']
                # 如果是雷達回波圖相關的文字
                if text == '雷達回波圖' or text == '雷達回波':
                    # 傳送雷達回波圖 ( 加上時間戳記 )
                    reply_image(f'https://cwbopendata.s3.ap-northeast-
```

```
1.amazonaws.com/MSC/O-A0058-003.png?{time.time_ns()}',
reply_token, access_token)
            else:
                # 如果是一般文字，直接回覆同樣的文字
                reply_message(text, reply_token, access_token)
    except:
        print('error')
    return 'OK'

if __name__ == "__main__":
    # Colab 使用，本機環境請刪除
    run_with_ngrok(app)
    app.run()

# LINE 回傳圖片函式
def reply_image(msg, rk, token):
    headers = {'Authorization':f'Bearer {token}','Content-
Type':'application/json'}
    body = {
    'replyToken':rk,
    'messages':[{
        'type': 'image',
        'originalContentUrl': msg,
        'previewImageUrl': msg
    }]
    }
    req = requests.request('POST', 'https://api.line.me/v2/bot/
message/reply', headers=headers,data=json.dumps(body).encode('utf-8'))
    print(req.text)
```
(範例程式碼：ch7/code02.py)

完成後重新執行 (使用 Colab + ngrok 需要重新設定 Webhook)，在 LINE 裡面輸入「雷達回波圖」或「雷達回波」，就會收到雷達回波圖的圖片。

回傳地震資訊

如果要從中央氣象局資料開放平臺取得地震資訊資料，就必須註冊並登入「氣象資料開放平臺」，註冊成功後會成為「一般會員」，點擊「取得授權碼」按鈕，會出現個人的授權碼，如果授權碼被盜用或出現問題，可點擊「更新授權碼」重新產生。

> 氣象資料開放平臺：
>
> https://opendata.cwb.gov.tw/index

API授權碼

本平臺提供透過URL下載檔案以及 RESTful API 資料擷取方法取用資料，惟因本平臺採用會員服務機制，需帶入資料項目代碼以及有效會員之授權碼，方可取得各式開放資料。其中，資料項目代碼可至資料清單列表查詢。

一、取得授權碼

會員之授權碼可於下方按鈕取得

取得授權碼　　□□□□□□□□□□□□□□□□

二、更新授權碼

一旦更新授權碼後，舊的授權碼將永久失效，並且更新授權碼後七日內無法再進行更新。

更新授權碼

點擊「資料主題」，選擇「天氣警特報」，搜尋「地震」，找到「小區域有感地震報告」資料集。

小區域有感地震報告：
https://opendata.cwb.gov.tw/dataset/warning/E-A0016-001

搜尋結果

🔍 地震

搜尋結果	
資料名稱	**資料編號**
顯著有感地震報告資料-顯著有感地震報告　ZIP　API	E-A0015-001
顯著有感地震報告資料-顯著有感地震報告(英文)　ZIP　API	E-A0015-002
小區域有感地震報告資料-小區域有感地震報告　ZIP　API	E-A0016-001
小區域有感地震報告資料-小區域有感地震報告(英文)　ZIP　API	E-A0016-002

點擊 API 連結，開啟「中央氣象局開放資料平臺之資料擷取 API」的頁面，
輸入個人的授權碼。

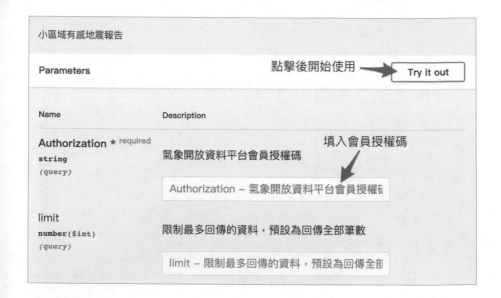

輸入後點擊下方的「Execute」，產生地震資訊的 JSON 連結。

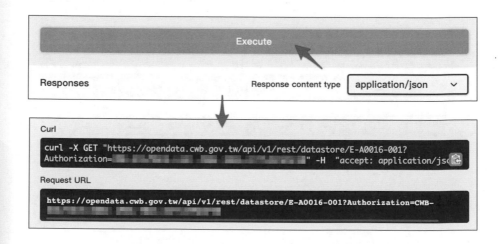

用瀏覽器開啟地震資訊，就能觀察 JSON 檔案的結構。

```
▼ object {3}
      success : true
   ▼ result {2}
         resource_id : E-A0016-001
      ▶ fields [21]
   ▼ records {2}
         datasetDescription : 地震報告
      ▼ earthquake [16]
         ▼ 0 {9}
               earthquakeNo : 111000
               reportType : 地震報告
               reportColor : 綠色
               reportContent : 03/12-06:25花蓮縣秀林鄉發生規模3.3有感地震，最大震度花蓮縣銅門2級。
               reportImageURI : https://scweb.cwb.gov.tw/webdata/OLDEQ/202203/2022031206252133_H.png
               reportRemark : 本報告係中央氣象局地震觀測網即時地震資料地震速報之結果。
               web : https://scweb.cwb.gov.tw/zh-tw/earthquake/details/2022031206252133
            ▼ earthquakeInfo {5}
                  originTime : 2022-03-12 06:25:21
                  source : 中央氣象局
               ▼ depth {2}
                     value : 19.4
                     unit : 公里
               ▼ epiCenter {3}
                     location : 花蓮縣政府西南西方  20.0  公里（位於花蓮縣秀林鄉）
                  ▼ epiCenterLat {2}
                        value : 23.9
                        unit : 度
                  ▼ epiCenterLon {2}
                        unit : 度
                        value : 121.45
               ▼ magnitude {2}
                     magnitudeType : 芮氏規模
                     magnitudeValue : 3.3
```

使用 Requests 函式庫的 get 的方法，抓取氣象觀測資料的 JSON 網址，接著使用字典的取值方法，搭配 for 迴圈，就能印出城市名稱、區域名稱和觀測點名稱。

```python
import requests
url = '你取得的地震資訊 JSON 網址'
data = requests.get(url)
data_json = data.json()
eq = data_json['records']['earthquake']    # 轉換成 json 格式
for i in eq:
    loc = i['earthquakeInfo']['epiCenter']['location']          # 地震地點
    val = i['earthquakeInfo']['magnitude']['magnitudeValue']   # 芮氏規模
    dep = i['earthquakeInfo']['depth']['value']                # 地震深度
    eq_time = i['earthquakeInfo']['originTime']                # 地震時間
    print(f'地震發生於 {loc}，芮氏規模 {val} 級，深度 {dep} 公里，發生時間
{eq_time}')
```

（範例程式碼：ch7/code04.py）

```
地震發生於花蓮縣政府西南西方  20.0  公里 (位於花蓮縣秀林鄉), 芮氏規模 3.3 級, 深度 19.4 公里, 發生時間 2022-03-12 06:25:21
地震發生於花蓮縣政府東方  135.7  公里 (位於臺灣東部海域), 芮氏規模 5.0 級, 深度 73.2 公里, 發生時間 2022-03-08 02:05:55
地震發生於宜蘭縣政府東方  48.6  公里 (位於臺灣東部海域), 芮氏規模 5.1 級, 深度 109.2 公里, 發生時間 2022-03-06 11:51:56
地震發生於花蓮縣政府西南西方  14.8  公里 (位於花蓮縣秀林鄉), 芮氏規模 3.5 級, 深度 19.8 公里, 發生時間 2022-03-06 09:24:17
地震發生於宜蘭縣政府南南東方  42.3  公里 (位於臺灣東部海域), 芮氏規模 4.0 級, 深度 32.0 公里, 發生時間 2022-03-05 14:54:08
地震發生於臺東縣政府北北西方  31.1  公里 (位於臺東縣海端鄉), 芮氏規模 3.5 級, 深度 5.0 公里, 發生時間 2022-03-04 15:54:46
地震發生於宜蘭縣政府南南東方  28.9  公里 (位於宜蘭縣近海), 芮氏規模 3.7 級, 深度 11.4 公里, 發生時間 2022-03-03 19:33:22
地震發生於嘉義縣政府南西方  14.2  公里 (位於嘉義縣義竹鄉), 芮氏規模 2.8 級, 深度 9.6 公里, 發生時間 2022-03-03 01:56:05
地震發生於花蓮縣政府南南西方  73.5  公里 (位於花蓮縣卓溪鄉), 芮氏規模 3.5 級, 深度 11.7 公里, 發生時間 2022-03-02 20:31:48
地震發生於臺南市政府東南方  9.9  公里 (位於臺南市仁德區), 芮氏規模 3.3 級, 深度 13.5 公里, 發生時間 2022-03-01 18:27:53
地震發生於花蓮縣政府北北東方  30.9  公里 (位於臺灣東部海域), 芮氏規模 4.3 級, 深度 61.4 公里, 發生時間 2022-02-26 00:42:24
地震發生於花蓮縣政府東方  82.6  公里 (位於臺灣東部海域), 芮氏規模 4.9 級, 深度 25.9 公里, 發生時間 2022-02-20 18:49:44
地震發生於花蓮縣政府東北東方  33.2  公里 (位於臺灣東部海域), 芮氏規模 3.9 級, 深度 11.7 公里, 發生時間 2022-02-20 14:13:49
地震發生於宜蘭縣政府東北方  16.9  公里 (位於臺灣東部海域), 芮氏規模 4.0 級, 深度 48.5 公里, 發生時間 2022-02-18 22:27:20
地震發生於宜蘭縣政府南南東方  33.3  公里 (位於宜蘭縣近海), 芮氏規模 3.6 級, 深度 45.1 公里, 發生時間 2022-02-18 10:05:27
地震發生於宜蘭縣政府南方  34.3  公里 (位於宜蘭縣南澳鄉), 芮氏規模 3.7 級, 深度 32.4 公里, 發生時間 2022-02-16 11:35:38
```

能夠取得地震資訊後,修改剛剛的程式碼,加入抓取地震資訊的函式,修改的重點如下 (詳細說明寫在程式碼的註解中):

- 建立一個 reply_message 函式,使用 requests 的 GET 方法傳送訊息。

- 建立一個 push_message 函式,使用 requests 的 GET 方法傳送訊息 (因為 reply 的方法只能傳送一次,如果要傳送兩次就需要使用 push 的方法)。

- 建立一個 earth_quake 函式,負責爬取地震資訊,組合成串列,串列的第一個項目為文字說明,第二個項目為地震圖,爬取資訊後,回傳第一筆資料。

- 使用 if 判斷訊息是否出現「地震」或「地震資訊」。

```python
# Colab 使用,本機環境請刪除
from flask_ngrok import run_with_ngrok

from flask import Flask, request
from linebot import LineBotApi, WebhookHandler
from linebot.models import MessageEvent, TextMessage, TextSendMessage
import requests, json, time


app = Flask(__name__)

access_token = '你的 LINE Channel access token'
channel_secret = '你的 LINE Channel secret'


@app.route("/", methods=['POST'])
```

```python
def linebot():
    body = request.get_data(as_text=True)
    try:
        line_bot_api = LineBotApi(access_token)
        handler = WebhookHandler(channel_secret)
        signature = request.headers['X-Line-Signature']
        handler.handle(body, signature)
        json_data = json.loads(body)
        reply_token = json_data['events'][0]['replyToken']
        user_id = json_data['events'][0]['source']['userId']
        print(json_data)
        if 'message' in json_data['events'][0]:
            if json_data['events'][0]['message']['type'] == 'text':
                text = json_data['events'][0]['message']['text']
                if text == '雷達回波圖' or text == '雷達回波':
                    reply_image(f'https://cwbopendata.s3.ap-northeast-
1.amazonaws.com/MSC/O-A0058-003.png?{time.time_ns()}', reply_token,
access_token)
                # 如果是地震相關的文字
                elif text == '地震資訊' or text == '地震':
                    # 爬取地震資訊
                    msg = earth_quake()
                    # 傳送地震資訊 ( 用 push 方法，因為 reply 只能用一次 )
                    push_message(msg[0], user_id, access_tokcn)
                    # 傳送地震圖片 ( 用 reply 方法 )
                    reply_image(msg[1], reply_token, access_token)
                else:
                    # 如果是一般文字，直接回覆同樣的文字
                    reply_message(text, reply_token, access_token)
    except:
        print('error')
    return 'OK'

if __name__ == "__main__":
    # Colab 使用，本機環境請刪除
    run_with_ngrok(app)
    app.run()

# 地震資訊函式
```

```python
def earth_quake():
    # 預設回傳的訊息
    msg = [' 找不到地震資訊 ','https://example.com/demo.jpg']
    try:
        code = ' 你的氣象資料授權碼 '
        url = f'https://opendata.cwb.gov.tw/api/v1/rest/datastore/
E-A0016-001?Authorization={code}'
        # 爬取地震資訊網址
        e_data = requests.get(url)
        # json 格式化訊息內容
        e_data_json = e_data.json()
        # 取出地震資訊
        eq = e_data_json['records']['earthquake']
        for i in eq:
            loc = i['earthquakeInfo']['epiCenter']['location']# 地震地點
            val = i['earthquakeInfo']['magnitude']['magnitudeValue']
# 地震規模
            dep = i['earthquakeInfo']['depth']['value']        # 地震深度
            eq_time = i['earthquakeInfo']['originTime']         # 地震時間
            img = i['reportImageURI']                           # 地震圖
            msg = [f'{loc} ,芮氏規模 {val} 級,深度 {dep} 公里,發生時間 {eq_
time} 。', img]
            break        # 取出第一筆資料後就 break
        return msg       # 回傳 msg
    except:
        return msg       # 如果取資料有發生錯誤,直接回傳 msg


# LINE push 訊息函式
def push_message(msg, uid, token):
    headers = {'Authorization':f'Bearer {token}','Content-
Type':'application/json'}
    body = {
    'to':uid,
    'messages':[{
            "type": "text",
            "text": msg
        }]
    }
    req = requests.request('POST', 'https://api.line.me/v2/bot/
```

```
message/push', headers=headers,data=json.dumps(body).encode('utf-8'))
    print(req.text)

# LINE 回傳訊息函式
def reply_message(msg, rk, token):
    headers = {'Authorization':f'Bearer {token}','Content-
Type':'application/json'}
    body = {
    'replyToken':rk,
    'messages':[{
            "type": "text",
            "text": msg
        }]
    }
    req = requests.request('POST', 'https://api.line.me/v2/bot/
message/reply', headers=headers,data=json.dumps(body).encode('utf-8'))
    print(req.text)

# LINE 回傳圖片函式
def reply_image(msg, rk, token):
    headers = {'Authorization':f'Bearer {token}','Content-
Type':'application/json'}
    body = {
    'replyToken':rk,
    'messages':[{
            'type': 'image',
            'originalContentUrl': msg,
            'previewImageUrl': msg
        }]
    }
    req = requests.request('POST', 'https://api.line.me/v2/bot/
message/reply', headers=headers,data=json.dumps(body).encode('utf-8'))
    print(req.text)
```

(範例程式碼：ch7/code04.py)

完成後執行程式（使用 Colab + ngrok 需要重新設定 Webhook），LINE 裡面輸入「地震」或「地震資訊」，就會收到最新一筆的地震資料和地震圖。

7-2 氣象機器人 (2) - 目前氣象資訊

只是透過聊天文字回傳氣象資訊還不夠，接下來會串接目前氣象資訊，實作透過 LINE 預設的功能提供地址，就能回傳該地址的即時天氣資訊。

取得地址資訊

修改剛剛的程式範例，在「if 'message' in json_data['events'][0]:」的判斷式裡面加上 message type 為 location 的判斷，如果是 location，就取出 address 的內容，並使用 replace 將「台」換成「臺」(因為氣象局都是用「臺」，但 LINE 的 location 都用「台」)

```
# Colab 使用，本機環境請刪除
from flask_ngrok import run_with_ngrok

from flask import Flask, request
from linebot import LineBotApi, WebhookHandler
from linebot.models import MessageEvent, TextMessage, TextSendMessage
import requests, json, time

app = Flask(__name__)

access_token = '你的 LINE Channel access token'
channel_secret = '你的 LINE Channel secret'

@app.route("/", methods=['POST'])
def linebot():
    body = request.get_data(as_text=True)
    try:
        line_bot_api = LineBotApi(access_token)
        handler = WebhookHandler(channel_secret)
        signature = request.headers['X-Line-Signature']
        handler.handle(body, signature)
        json_data = json.loads(body)
```

```python
        reply_token = json_data['events'][0]['replyToken']
        user_id = json_data['events'][0]['source']['userId']
        print(json_data)
        if 'message' in json_data['events'][0]:
            # 如果 message 的類型是地圖 location
            if json_data['events'][0]['message']['type'] =='location':
                # 取出地址資訊，並將「台」換成「臺」
                address = json_data['events'][0]['message']
['address'].replace('台','臺')
                print(address)
            if json_data['events'][0]['message']['type'] == 'text':
                text = json_data['events'][0]['message']['text']
                if text == '雷達回波圖' or text == '雷達回波':
                    reply_image(f'https://cwbopendata.s3.ap-northeast-
1.amazonaws.com/MSC/O-A0058-003.png?{time.time_ns()}',
reply_token, access_token)
                elif text == '地震資訊' or text == '地震':
                    msg = earth_quake()
                    push_message(msg[0], user_id, access_token)
                    reply_image(msg[1], reply_token, access_token)
                else:
                    reply_message(text, reply_token, access_token)
    except:
        print('error')
    return 'OK'

if __name__ == "__main__":
    # Colab 使用，本機環境請刪除
    run_with_ngrok(app)
    app.run()


# 地震資訊函式
def earth_quake():
    msg = [' 找不到地震資訊 ','https://example.com/demo.jpg']
    try:
        code = ' 你的氣象資料授權碼 '
        url = f'https://opendata.cwb.gov.tw/api/v1/rest/datastore/
E-A0016-001?Authorization={code}'
        # 爬取地震資訊網址
```

```python
        e_data = requests.get(url)
        # json 格式化訊息內容
        e_data_json = e_data.json()
        # 取出地震資訊
        eq = e_data_json['records']['earthquake']
        for i in eq:
            loc = i['earthquakeInfo']['epiCenter']['location']
            val = i['earthquakeInfo']['magnitude']['magnitudeValue']
            dep = i['earthquakeInfo']['depth']['value']
            eq_time = i['earthquakeInfo']['originTime']
            img = i['reportImageURI']
            msg = [f'{loc},芮氏規模 {val} 級,深度 {dep} 公里,發生時間 {eq_
time}。', img]
            break
        return msg
    except:
        return msg


# LINE push 訊息函式
def push_message(msg, uid, token):
    headers = {'Authorization':f'Bearer {token}','Content-
Type':'application/json'}
    body = {
    'to':uid,
    'messages':[{
            "type": "text",
            "text": msg
        }]
    }
    req = requests.request('POST', 'https://api.line.me/v2/bot/
message/push', headers=headers,data=json.dumps(body).encode('utf-8'))
    print(req.text)


# LINE 回傳訊息函式
def reply_message(msg, rk, token):
    headers = {'Authorization':f'Bearer {token}','Content-
Type':'application/json'}
    body = {
    'replyToken':rk,
```

```
    'messages':[{
            "type": "text",
            "text": msg
        }]
    }
    req = requests.request('POST', 'https://api.line.me/v2/bot/
message/reply', headers=headers,data=json.dumps(body).encode('utf-8'))
    print(req.text)


# LINE 回傳圖片函式
def reply_image(msg, rk, token):
    headers = {'Authorization':f'Bearer {token}','Content-
Type':'application/json'}
    body = {
    'replyToken':rk,
    'messages':[{
            'type': 'image',
            'originalContentUrl': msg,
            'previewImageUrl': msg
        }]
    }
    req = requests.request('POST', 'https://api.line.me/v2/bot/
message/reply', headers=headers,data=json.dumps(body).encode('utf-8'))
    print(req.text)
```
(範例程式碼：ch7/code05.py)

完成後執行程式 (使用 Colab + ngrok 需要重新設定 Webhook)，LINE 裡面
傳送地址資訊，在城市後台就會看到出現地址的內容。

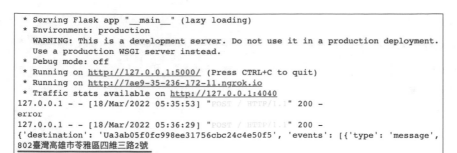

```
* Serving Flask app "__main__" (lazy loading)
* Environment: production
  WARNING: This is a development server. Do not use it in a production deployment.
  Use a production WSGI server instead.
* Debug mode: off
* Running on http://127.0.0.1:5000/ (Press CTRL+C to quit)
* Running on http://7ae9-35-236-172-11.ngrok.io
* Traffic stats available on http://127.0.0.1:4040
127.0.0.1 - - [18/Mar/2022 05:35:53] "POST / HTTP/1.1" 200 -
error
127.0.0.1 - - [18/Mar/2022 05:36:29] "POST / HTTP/1.1" 200 -
{'destination': 'Ua3ab05f0fc998ee31756cbc24c4e50f5', 'events': [{'type': 'message',
802臺灣高雄市苓雅區四維三路2號
```

🖐 回傳目前氣象資訊

註冊並登入「氣象資料開放平臺」，點擊「資料主題」，選擇「觀測」，找到「自動氣象站 - 氣象觀測資料」資料集。

從氣象資料開放平臺裡，找到「自動氣象站-氣象觀測資料」和「局屬氣象站-現在天氣觀測報告」兩個資料集，複製 JSON 網址（「自動氣象站-氣象觀測資料」包含鄉鎮區域的資訊，「局屬氣象站-現在天氣觀測報告」則是以縣市為主的天氣資訊）。

資料連結（需要登入才看得到 JSON 連結）

● 自動氣象站 - 氣象觀測資料
   https://opendata.cwb.gov.tw/dataset/observation/O-A0001-001

● 局屬氣象站 - 現在天氣觀測報告
   https://opendata.cwb.gov.tw/dataset/observation/O-A0003-001

不論是用 JSON 檔案還是 API 的方式，開啟氣象觀測資料後，可從 JSON 物件裡找到地點的名稱、城市名稱等觀測點資訊，weatherElement 裡則是該地點的現在氣象觀測資料，查看氣象局的 pdf 說明，可了解 weatherElement 裡觀測因子所代表的意義。

氣象局說明資料：

https://opendata.cwb.gov.tw/opendatadoc/DIV2/A0001-001.pdf

```
▼ object {1}
  ▼ cwbopendata {10}
        @xmlns : urn:cwb:gov:tw:cwbcommon:0.1
        identifier : 54cee551-0ff9-476a-9858-2a53ed2a09c2
        sender : weather@cwb.gov.tw
        sent : 2022-03-14T09:28:05+08:00
        status : Actual
        msgType : Issue
        dataid : CWB_A0001
        scope : Public
        dataset : null
     ▼ location [450]
        ▼ 0 {9}
              lat : 25.098133
              lon : 121.508275
              lat_wgs84 : 25.0963555555556
              lon_wgs84 : 121.516505555556
              locationName : 科教館
              stationId : C0A770
           ▶ time {1}
           ▶ weatherElement [14]
           ▼ parameter [4]
              ▼ 0 {2}
                    parameterName : CITY
                    parameterValue : 臺北市
              ▼ 1 {2}
                    parameterName : CITY_SN
                    parameterValue : 01
              ▼ 2 {2}
                    parameterName : TOWN
                    parameterValue : 士林區
```

修改 Python 程式，新增「**current_weather**」函式，函式包含一個 **address** 參數，負責承接 **LINE** 收到的地址資訊，並根據地址資訊查詢目前即時天氣，完成後在上方新增一個 **reply_message** 負責回覆訊息，程式相關重點如下 (詳細說明寫在程式碼的註解中)：

- 定義爬取資料的函式 get_data，篩選出 JSON 資料裡的縣市名稱、鄉鎮行政區、氣溫、相對濕度和累積雨量。

- 由於可能會遇到找不到行政區的狀況，所以會額外獨立縣市名稱作為其中一個 key。

- 如果是以縣市名稱為 key，則內容是採用該縣市裡鄉鎮行政區的平均數值（例如新北市有十個區，每個區有各自的溫度，新北市的溫度就是這十個區的平均）。

- 計算串列裡的平均數使用標準函式庫的 statistics.mean 方法。

- 定義 check_data 函式，處理小於 0 的數值（因有時數據會出現 -99 之類的數值，所以將小於 0 的數值回傳為 False）。

- 定義 msg_content 函式，產生回傳的訊息。

```python
# Colab 使用，本機環境請刪除
from flask_ngrok import run_with_ngrok

from flask import Flask, request
from linebot import LineBotApi, WebhookHandler
from linebot.models import MessageEvent, TextMessage, TextSendMessage
import requests, json, time, statistics  # import statistics 函式庫

app = Flask(__name__)

access_token = '你的 LINE Channel access token'
channel_secret = '你的 LINE Channel secret'

@app.route("/", methods=['POST'])
def linebot():
    body = request.get_data(as_text=True)
    try:
        line_bot_api = LineBotApi(access_token)
        handler = WebhookHandler(channel_secret)
        signature = request.headers['X-Line-Signature']
        handler.handle(body, signature)
```

```python
        json_data = json.loads(body)
        reply_token = json_data['events'][0]['replyToken']
        user_id = json_data['events'][0]['source']['userId']
        print(json_data)
        if 'message' in json_data['events'][0]:
            if json_data['events'][0]['message']['type'] =='location':
                # 取出地址資訊，並將「台」換成「臺」
                address = json_data['events'][0]['message']
['address'].replace('台','臺')
                print(address)
                # 回覆爬取到的相關氣象資訊
                reply_message(f'{address}\n\n{current_
weather(address)}', reply_token, access_token)
            if json_data['events'][0]['message']['type'] == 'text':
                text = json_data['events'][0]['message']['text']
                if text == '雷達回波圖' or text == '雷達回波':
                    reply_image(f'https://cwbopendata.s3.ap-northeast-1.
amazonaws.com/MSC/O-A0058-003.png?{time.time_ns()}', reply_token,
access_token)                    elif text == '地震資訊' or text == '地震':
                    msg = earth_quake()
                    push_message(msg[0], user_id, access_token)
                    reply_image(msg[1], reply_token, access_token)
                else:
                    reply_message(text, reply_token, access_token)
    except:
        print('error')
    return 'OK'


if __name__ == "__main__":
    # Colab 使用，本機環境請刪除
    run_with_ngrok(app)
    app.run()


# 地震資訊函式
def earth_quake():
    # 預設回傳的訊息
    msg = ['找不到地震資訊','https://example.com/demo.jpg']
    try:
        code = '你的氣象資料授權碼'
```

```python
        url = f'https://opendata.cwb.gov.tw/api/v1/rest/datastore/
E-A0016-001?Authorization={code}'
        e_data = requests.get(url)
        e_data_json = e_data.json()
        eq = e_data_json['records']['earthquake']
        for i in eq:
            loc = i['earthquakeInfo']['epiCenter']['location']
            val = i['earthquakeInfo']['magnitude']['magnitudeValue']
            dep = i['earthquakeInfo']['depth']['value']
            eq_time = i['earthquakeInfo']['originTime']
            img = i['reportImageURI']
            msg = [f'{loc}，芮氏規模 {val} 級，深度 {dep} 公里，發生時間 {eq_
time}。', img]
            break
        return msg
    except:
        return msg


# LINE push 訊息函式
def push_message(msg, uid, token):
    headers = {'Authorization':f'Bearer {token}','Content-
Type':'application/json'}
    body = {
    'to':uid,
    'messages':[{
            "type": "text",
            "text": msg
        }]
    }
    req = requests.request('POST', 'https://api.line.me/v2/bot/
message/push', headers=headers,data=json.dumps(body).encode('utf-8'))
    print(req.text)


# LINE 回傳訊息函式
def reply_message(msg, rk, token):
    headers = {'Authorization':f'Bearer {token}','Content-
Type':'application/json'}
    body = {
    'replyToken':rk,
```

```
    'messages':[{
          "type": "text",
          "text": msg
      }]
    }
    req = requests.request('POST', 'https://api.line.me/v2/bot/
message/reply', headers=headers,data=json.dumps(body).encode('utf-8'))
    print(req.text)

# LINE 回傳圖片函式
def reply_image(msg, rk, token):
    headers = {'Authorization':f'Bearer {token}','Content-
Type':'application/json'}
    body = {
    'replyToken':rk,
    'messages':[{
          'type': 'image',
          'originalContentUrl': msg,
          'previewImageUrl': msg
      }]
    }
    req = requests.request('POST', 'https://api.line.me/v2/bot/
message/reply', headers=headers,data=json.dumps(body).encode('utf-8'))
    print(req.text)

# 目前天氣函式
def current_weather(address):
    # 定義好待會要用的變數
    city_list, area_list, area_list2 = {}, {}, {}
    # 預設回傳訊息
    msg = '找不到氣象資訊。'

    # 定義取得資料的函式
    def get_data(url):
        # 爬取目前天氣網址的資料
        w_data = requests.get(url)
        # json 格式化訊息內容
        w_data_json = w_data.json()
        # 取出對應地點的內容
```

```
        location = w_data_json['cwbopendata']['location']
        for i in location:
            name = i['locationName']                        # 測站地點
            city = i['parameter'][0]['parameterValue']      # 縣市名稱
            area = i['parameter'][2]['parameterValue']      # 鄉鎮行政區
            temp = check_data(i['weatherElement'][3]['elementValue']
['value'])                          # 氣溫
            humd = check_data(round(float(i['weatherElement'][4]
['elementValue']['value'] )*100 ,1)) # 相對濕度
            r24 = check_data(i['weatherElement'][6]['elementValue']
['value'])                          # 累積雨量
            if area not in area_list:
                # 以鄉鎮區域為 key，儲存需要的資訊
                area_list[area] = {'temp':temp, 'humd':humd, 'r24':r24}
            if city not in city_list:
                # 以主要縣市名稱為 key，準備紀錄裡面所有鄉鎮的數值
                city_list[city] = {'temp':[], 'humd':[], 'r24':[]}
            city_list[city]['temp'].append(temp)    # 記錄主要縣市裡鄉鎮區
域的溫度（串列格式）
            city_list[city]['humd'].append(humd)    # 記錄主要縣市裡鄉鎮區
域的濕度（串列格式）
            city_list[city]['r24'].append(r24)      # 記錄主要縣市裡鄉鎮區
域的雨量（串列格式）

    # 定義如果數值小於 0，回傳 False 的函式
    def check_data(e):
        return False if float(e)<0 else float(e)

    # 定義產生回傳訊息的函式
    def msg_content(loc, msg):
        a = msg
        for i in loc:
            if i in address: # 如果地址裡存在 key 的名稱
                temp = f"氣溫 {loc[i]['temp']} 度，" if loc[i]['temp']
!= False else ''
                humd = f"相對濕度 {loc[i]['humd']}%，" if loc[i]['humd']
!= False else ''
                r24 = f"累積雨量 {loc[i]['r24']}mm" if loc[i]['r24'] !=
False else ''
```

```
            description = f'{temp}{humd}{r24}'.strip('')
            a = f'{description}。' # 取出 key 的內容作為回傳訊息使用
            break
    return a

try:
    # 因為目前天氣有兩組網址，兩組都爬取
    code = '你的氣象資料授權碼'
    get_data(f'https://opendata.cwb.gov.tw/fileapi/v1/opendataapi/
O-A0001-001?Authorization={code}&downloadType=WEB&format=JSON')
    get_data(f'https://opendata.cwb.gov.tw/fileapi/v1/opendataapi/
O-A0003-001?Authorization={code}&downloadType=WEB&format=JSON')

    for i in city_list:
        if i not in area_list2:
            # 將主要縣市裡的數值平均後，以主要縣市名稱為 key，再度儲存一次，
如果找不到鄉鎮區域，就使用平均數值
            area_list2[i] = {'temp':round(statistics.mean(city_
list[i]['temp']),1),
                             'humd':round(statistics.mean(city_
list[i]['humd']),1),
                             'r24':round(statistics.mean(city_
list[i]['r24']),1)
                            }
    msg = msg_content(area_list2, msg)  # 將訊息改為「大縣市」
    msg = msg_content(area_list, msg)   # 將訊息改為「鄉鎮區域」
    return msg    # 回傳 msg
except:
    return msg    # 如果取資料有發生錯誤，直接回傳 msg
```

(範例程式碼：ch7/code06.py)

完成後執行程式（使用 Colab + ngrok 需要重新設定 Webhook），LINE 裡面
傳送地址資訊，就會回覆目前所在位置的即時氣象資訊。

7-3　氣象機器人 (3) - 天氣預報和空氣品質

目前的氣象機器人已經可以串接雷達回波圖、地震資訊和目前天氣，接下來要串接天氣預報和空氣品質資訊，同樣使用 LINE 預設的功能提供地址，就能回傳該地址的天氣預報和空氣品質資訊。

回傳天氣預報資訊

註冊並登入「氣象資料開放平臺」，找到「一般天氣預報 - 今明 36 小時天氣預報」資料集複製 JSON 網址。

> **一般天氣預報 - 今明 36 小時天氣預報**（需要登入才看得到 JSON 連結）：
> https://opendata.cwb.gov.tw/dataset/forecast/F-C0032-001

為了進一步取得鄉鎮行政區的天氣預報資料，再從氣象資料開放平臺的「預報」分類中，搜尋「2 天天氣」，可以看到每個資料集的代碼，屆時會透過這些代碼，取得對應的 JSON 檔案 (除了代碼，其他網址都相同)

搜尋結果	
資料名稱	資料編號
鄉鎮天氣預報-單一鄉鎮市區預報資料-宜蘭縣未來2天天氣預報 JSON　XML　API	F-D0047-001
鄉鎮天氣預報-單一鄉鎮市區預報資料-桃園市未來2天天氣預報 JSON　XML　API	F-D0047-005
鄉鎮天氣預報-單一鄉鎮市區預報資料-新竹縣未來2天天氣預報 JSON　XML　API	F-D0047-009

（ 2天天氣 搜尋框）

不論是用 JSON 檔案還是 API 的方式，開啟氣象預報資料後，可看見如下圖的 JSON 物件結構，weatherElement 裡的資料就是所需的天氣預報資料。

weatherElement 裡有五種的預報因子，可透過程式單純取出所需的預報因子。

預報因子	說明
Wx	天氣現象
MaxT	最高溫度
MinT	最低溫度
CI	舒適度
PoP	降雨機率

修改剛剛的 Python 程式，**新增「forecast 函式」**，**函式內容包含一個 address 參數，負責承接 LINE 收到的地址資訊，並根據地址資訊查詢氣象預報，接著修改收到地址資訊後的 reply 訊息內容**，程式相關重點如下（詳細說明寫在程式碼的註解中）：

● 設定一個變數 json_api，用字典的方式儲存鄉鎮行政區的代碼，key 是縣市名稱。

● 先抓取主要縣市的氣象預報資料，抓到資料後組合成文字，取代原本 msg 變數的內容。

● 再抓取鄉鎮行政區的氣象預報資料，如果沒有資料就可以使用縣市預報資料。

```python
# Colab 使用，本機環境請刪除
from flask_ngrok import run_with_ngrok

from flask import Flask, request
from linebot import LineBotApi, WebhookHandler
from linebot.models import MessageEvent, TextMessage, TextSendMessage
import requests, json, time, statistics

app = Flask(__name__)

access_token = '你的 LINE Channel access token'
channel_secret = '你的 LINE Channel secret'

@app.route("/", methods=['POST'])
def linebot():
    body = request.get_data(as_text=True)
    try:
        line_bot_api = LineBotApi(access_token)
        handler = WebhookHandler(channel_secret)
        signature = request.headers['X-Line-Signature']
        handler.handle(body, signature)
        json_data = json.loads(body)
        reply_token = json_data['events'][0]['replyToken']
```

```
            user_id = json_data['events'][0]['source']['userId']
        print(json_data)
        if 'message' in json_data['events'][0]:
            if json_data['events'][0]['message']['type'] == 'location':
                address = json_data['events'][0]['message']
['address'].replace('台','臺')
                print(address)
                # 修改回覆訊息，加入天氣預報內容
                reply_message(f'{address}\n\n{current_
weather(address)}\n\n{forecast(address)}', reply_token, access_token)
            if json_data['events'][0]['message']['type'] == 'text':
                text = json_data['events'][0]['message']['text']
                if text == '雷達回波圖' or text == '雷達回波':
                    reply_image(f'https://cwbopendata.s3.ap-northeast-1.
amazonaws.com/MSC/O-A0058-003.png?{time.time_ns()}', reply_token,
access_token)
                elif text == '地震資訊' or text == '地震':
                    msg = earth_quake()
                    push_message(msg[0], user_id, access_token)
                    reply_image(msg[1], reply_token, access_token)
                else:
                    reply_message(text, reply_token, access_token)
    except:
        print('error')
    return 'OK'

if __name__ == "__main__":
    # Colab 使用，本機環境請刪除
    run_with_ngrok(app)
    app.run()

# 地震資訊函式
def earth_quake():
    msg = [' 找不到地震資訊 ','https://example.com/demo.jpg']
    try:
        code = ' 你的氣象資料授權碼 '
        url = f'https://opendata.cwb.gov.tw/api/v1/rest/datastore/
E-A0016-001?Authorization={code}'
        e_data = requests.get(url)
```

```
        e_data_json = e_data.json()
        eq = e_data_json['records']['earthquake']
        for i in eq:
            loc = i['earthquakeInfo']['epiCenter']['location']
            val = i['earthquakeInfo']['magnitude']['magnitudeValue']
            dep = i['earthquakeInfo']['depth']['value']
            eq_time = i['earthquakeInfo']['originTime']
            img = i['reportImageURI']
            msg = [f'{loc}，芮氏規模 {val} 級，深度 {dep} 公里，發生時間 {eq_
time}。', img]
            break
        return msg
    except:
        return msg


# LINE push 訊息函式
def push_message(msg, uid, token):
    headers = {'Authorization':f'Bearer {token}','Content-
Type':'application/json'}
    body = {
    'to':uid,
    'messages':[{
            "type": "text",
            "text": msg
        }]
    }
    req = requests.request('POST', 'https://api.line.me/v2/bot/
message/push', headers=headers,data=json.dumps(body).encode('utf-8'))
    print(req.text)


# LINE 回傳訊息函式
def reply_message(msg, rk, token):
    headers = {'Authorization':f'Bearer {token}','Content-
Type':'application/json'}
    body = {
    'replyToken':rk,
    'messages':[{
            "type": "text",
            "text": msg
```

```
        }]
    }
    req = requests.request('POST', 'https://api.line.me/v2/bot/
message/reply', headers=headers,data=json.dumps(body).encode('utf-8'))
    print(req.text)

# LINE 回傳圖片函式
def reply_image(msg, rk, token):
    headers = {'Authorization':f'Bearer {token}','Content-
Type':'application/json'}
    body = {
    'replyToken':rk,
    'messages':[{
        'type': 'image',
        'originalContentUrl': msg,
        'previewImageUrl': msg
    }]
    }
    req = requests.request('POST', 'https://api.line.me/v2/bot/
message/reply', headers=headers,data=json.dumps(body).encode('utf-8'))
    print(req.text)

# 目前天氣函式
def current_weather(address):
    city_list, area_list, area_list2 = {}, {}, {}
    msg = '找不到氣象資訊。'

    def get_data(url):
        w_data = requests.get(url)
        w_data_json = w_data.json()
        location = w_data_json['cwbopendata']['location']
        for i in location:
            name = i['locationName']
            city = i['parameter'][0]['parameterValue']
            area = i['parameter'][2]['parameterValue']
            temp = check_data(i['weatherElement'][3]['elementValue']
['value'])
            humd = check_data(round(float(i['weatherElement'][4]
['elementValue']['value'] )*100 ,1))
```

```python
            r24 = check_data(i['weatherElement'][6]['elementValue']
['value'])
            if area not in area_list:
                area_list[area] = {'temp':temp, 'humd':humd,
'r24':r24}
            if city not in city_list:
                city_list[city] = {'temp':[], 'humd':[], 'r24':[]}
            city_list[city]['temp'].append(temp)
            city_list[city]['humd'].append(humd)
            city_list[city]['r24'].append(r24)

    def check_data(e):
        return False if float(e)<0 else float(e)

    def msg_content(loc, msg):
        a = msg
        for i in loc:
            if i in address:
                temp = f"氣溫 {loc[i]['temp']} 度，" if loc[i]['temp']
!= False else ''
                humd = f"相對濕度 {loc[i]['humd']}%，" if loc[i]['humd']
!= False else ''
                r24 = f"累積雨量 {loc[i]['r24']}mm" if loc[i]['r24'] !=
False else ''
                description = f'{temp}{humd}{r24}'.strip('，')
                a = f'{description}。'
                break
        return a

    try:
        code = '你的氣象資料授權碼'
        get_data(f'https://opendata.cwb.gov.tw/fileapi/v1/opendataapi/
O-A0001-001?Authorization={code}&downloadType=WEB&format=JSON')
        get_data(f'https://opendata.cwb.gov.tw/fileapi/v1/opendataapi/
O-A0003-001?Authorization={code}&downloadType=WEB&format=JSON')

        for i in city_list:
            if i not in area_list2:
                area_list2[i] = {'temp':round(statistics.mean(city_
```

```
list[i]['temp']),1),
                                'humd':round(statistics.mean(city_
list[i]['humd']),1),
                                'r24':round(statistics.mean(city_
list[i]['r24']),1)
                                }
        msg = msg_content(area_list2, msg)
        msg = msg_content(area_list, msg)
        return msg      # 回傳 msg
    except:
        return msg


# 氣象預報函式
def forecast(address):
    area_list = {}
    # 將主要縣市個別的 JSON 代碼列出
    json_api = {"宜蘭縣":"F-D0047-001","桃園市":"F-D0047-005","新竹縣
":"F-D0047-009","苗栗縣":"F-D0047-013",
            "彰化縣":"F-D0047-017","南投縣":"F-D0047-021","雲林縣":"F-
D0047-025","嘉義縣":"F-D0047-029",
            "屏東縣":"F-D0047-033","臺東縣":"F-D0047-037","花蓮縣":"F-
D0047-041","澎湖縣":"F-D0047-045",
            "基隆市":"F-D0047-049","新竹市":"F-D0047-053","嘉義市":"F-
D0047-057","臺北市":"F-D0047-061",
            "高雄市":"F-D0047-065","新北市":"F-D0047-069","臺中市":"F-
D0047-073","臺南市":"F-D0047-077",
            "連江縣":"F-D0047-081","金門縣":"F-D0047-085"}
    msg = '找不到天氣預報資訊。'      # 預設回傳訊息
    try:
        code = '你的氣象開放平台授權碼'
        url = f'https://opendata.cwb.gov.tw/fileapi/v1/opendataapi/
F-C0032-001?Authorization={code}&downloadType=WEB&format=JSON'
        f_data = requests.get(url)      # 取得主要縣市預報資料
        f_data_json = f_data.json()     # json 格式化訊息內容
        # 取得縣市的預報內容
        location = f_data_json['cwbopendata']['dataset']['location']
        for i in location:
            city = i['locationName']      # 縣市名稱
```

```
        wx8 = i['weatherElement'][0]['time'][0]['parameter']
['parameterName']   # 天氣現象
        mint8 = i['weatherElement'][1]['time'][0]['parameter']
['parameterName']  # 最低溫
        maxt8 = i['weatherElement'][2]['time'][0]['parameter']
['parameterName']  # 最高溫
        ci8 = i['weatherElement'][2]['time'][0]['parameter']
['parameterName']   # 舒適度
        pop8 = i['weatherElement'][2]['time'][0]['parameter']
['parameterName']  # 降雨機率
        # 組合成回傳的訊息，存在以縣市名稱為 key 的字典檔裡
        area_list[city] = f' 未來 8 小時 {wx8}，最高溫 {maxt8} 度，最
低溫 {mint8} 度，降雨機率 {pop8} %'
    for i in area_list:
        if i in address:          # 如果使用者的地址包含縣市名稱
            msg = area_list[i]  # 將 msg 換成對應的預報資訊
            # 將進一步的預報網址換成對應的預報網址
            url =
f'https://opendata.cwb.gov.tw/api/v1/rest/datastore/{json_api[i]}?
Authorization={code}&elementName=WeatherDescription'
            f_data = requests.get(url)  # 取得主要縣市裡各個區域鄉鎮的
氣象預報
            f_data_json = f_data.json() # json 格式化訊息內容
            # 取得預報內容
            location = f_data_json['records']['locations'][0]
['location']
            break
    for i in location:
        city = i['locationName']    # 取得縣市名稱
        wd = i['weatherElement'][0]['time'][1]['elementValue'][0]
['value']  # 綜合描述
        if city in address:          # 如果使用者的地址包含鄉鎮區域名稱
            msg = f' 未來八小時天氣 {wd}' # 將 msg 換成對應的預報資訊
            break
    return msg  # 回傳 msg
except:
    return msg  # 如果取資料有發生錯誤，直接回傳 msg
```

(範例程式碼：ch7/code07.py)

7-39

完成後執行程式 (使用 Colab + ngrok 需要重新設定 Webhook)，LINE 裡面
傳送地址資訊，除了會回覆目前所在位置的即時氣象，還會額外包含天氣預
報的資訊。

回傳空氣品質資訊

從政府資料開放平臺裡，搜尋「空氣品質指標」，就能開啟空氣品質指標 (
AQI) 的資料頁面，資料有提供 JSON 格式與 CSV 格式，接下來的程式碼會
使用 JSON 格式，點擊對應的按鈕，就可以開啟對應的 JSON API 內容。

- 政府資料開放平臺：https://data.gov.tw/
- 空氣品質指標 (AQI)：https://data.gov.tw/dataset/40448

啟 API 後，可以從內容架構裡，找到 records 的「鍵」，records 的內容由串列和字典（或稱物件和陣列）所組成，包含地點 SiteName、城市 County、AQI、PM2.5... 等空氣品質指標的數值。

```
{"info": {"notes": "", "label": "\u98a8\u5411(degrees)"}, "type": "text", "id": "WindDirec"},
"\u8cc7\u6599\u5efa\u7f6e\u65e5\u671f"}, "type": "text", "id": "PublishTime"}, {"info": {"not
"\u7d30\u61f8\u6d6e\u7c92\u79fb\u52d5\u5e73\u5747\u503c(\u03bcg/m3)"}, "type": "text",
"\u61f8\u6d6e\u7c92\u79fb\u52d5\u5e73\u5747\u503c(\u03bcg/m3)"}, "type": "text", "id":
"\u4e8c\u6c27\u5316\u786b\u79fb\u52d5\u5e73\u5747\u503c(ppb)"}, "type": "text", "id": "SO2_AV
"text", "id": "Longitude"}, {"info": {"notes": "", "label": "\u7def\u5ea6"}, "type": "text",
"\u6c2c\u7ad9\u7de8\u865f"}, "type": "text", "id": "SiteId"}], "__extras": {"api_key": "9be7b
"records": [{"SiteName":"基隆","County":"基隆市","AQI":"45","Pollutant":"","Status":"良
好","SO2":"0.6","CO":"0.18","CO_8hr":"0.2","O3":"49.8","O3_8hr":"49","PM10":"24","PM2.5":"9",
"PublishTime":"2021/12/08
14:00:00","PM2.5_AVG":"9","PM10_AVG":"28","SO2_AVG":"0","Longitude":"121.760056","Latitude":
{"SiteName":"汐止","County":"新北市","AQI":"31","Pollutant":"","Status":"良
好","SO2":"1","CO":"0.13","CO_8hr":"0.2","O3":"44.7","O3_8hr":"32","PM10":"20","PM2.5":"9","N
","PublishTime":"2021/12/08
14:00:00","PM2.5_AVG":"9","PM10_AVG":"25","SO2_AVG":"1","Longitude":"121.6423","Latitude":"25
{"SiteName":"萬里","County":"新北市","AQI":"82","Pollutant":"懸浮微粒","Status":"普
通","SO2":"0.7","CO":"0.16","CO_8hr":"0.1","O3":"52.5","O3_8hr":"53","PM10":"64","PM2.5":"11"
","PublishTime":"2021/12/08
14:00:00","PM2.5_AVG":"13","PM10_AVG":"82","SO2_AVG":"0","Longitude":"121.689881","Latitude":
14:58:01.217000"},{"SiteName":"新店","County":"新北市","AQI":"41","Pollutant":"","Status":"良
好","SO2":"0.5","CO":"0.19","CO_8hr":"0.2","O3":"50.3","O3_8hr":"44","PM10":"21","PM2.5":"9",
31","PublishTime":"2021/12/08
```

修改 Python 程式碼，**新增「aqi」函式，函式包含一個 address 參數，負責承接 LINE 收到的地址資訊，並透過地址資訊查詢出該地址相關的空氣品質資訊**，程式相關重點如下 (詳細說明寫在程式碼的註解中)：

● 先抓取主要縣市的空氣品質資料，抓到資料後組合成文字，取代原本 msg 變數的內容。

● 再抓取鄉鎮行政區的空氣品質資料，如果沒有資料就可以使用縣市空氣品質資料。

● 因為資料內容有些沒有空氣品質的描述，要自己用 if else 判斷後產生對應的文字。

```python
# Copyright © https://steam.oxxostudio.tw

# Colab 使用，本機環境請刪除
from flask_ngrok import run_with_ngrok

from flask import Flask, request
from linebot import LineBotApi, WebhookHandler
from linebot.models import MessageEvent, TextMessage, TextSendMessage
import requests, json, time, statistics

app = Flask(__name__)

access_token = '你的 LINE Channel access token'
channel_secret = '你的 LINE Channel secret'

@app.route("/", methods=['POST'])
def linebot():
    body = request.get_data(as_text=True)
    try:
        line_bot_api = LineBotApi(access_token)
        handler = WebhookHandler(channel_secret)
        signature = request.headers['X-Line-Signature']
        handler.handle(body, signature)
        json_data = json.loads(body)
        reply_token = json_data['events'][0]['replyToken']
```

```
        user_id = json_data['events'][0]['source']['userId']
        print(json_data)
        if 'message' in json_data['events'][0]:
            if json_data['events'][0]['message']['type'] =='location':
                address = json_data['events'][0]['message']
['address'].replace('台','臺')
                print(address)
                # 回覆爬取到的相關氣象資訊
                reply_message(f'{address}\n\n{current_
weather(address)}\n\n{aqi(address)}\n\n{forecast(address)}', reply_
token, access_token)
            if json_data['events'][0]['message']['type'] == 'text':
                text = json_data['events'][0]['message']['text']
                if text == '雷達回波圖' or text == '雷達回波':
                    reply_image(f'https://cwbopendata.s3.ap-northeast-1.
amazonaws.com/MSC/O-A0058-003.png?{time.time_ns()}', reply_token,
access_token)
                elif text == '地震資訊' or text == '地震':
                    msg = earth_quake()
                    push_message(msg[0], user_id, access_token)
                    reply_image(msg[1], reply_token, access_token)
                else:
                    reply_message(text, reply_token, access_token)
    except:
        print('error')
    return 'OK'


if __name__ == "__main__":
    # Colab 使用，本機環境請刪除
    run_with_ngrok(app)
    app.run()


# 地震資訊函式
def earth_quake():
    # 預設回傳的訊息
    msg = [' 找不到地震資訊 ','https://example.com/demo.jpg']
    try:
        code = ' 你的氣象資料授權碼 '
        url = f'https://opendata.cwb.gov.tw/api/v1/rest/datastore/
E-A0016-001?Authorization={code}'
```

```python
        e_data = requests.get(url)
        e_data_json = e_data.json()
        eq = e_data_json['records']['earthquake']
        for i in eq:
            loc = i['earthquakeInfo']['epiCenter']['location']
            val = i['earthquakeInfo']['magnitude']['magnitudeValue']
            dep = i['earthquakeInfo']['depth']['value']
            eq_time = i['earthquakeInfo']['originTime']
            img = i['reportImageURI']
            msg = [f'{loc}，芮氏規模 {val} 級，深度 {dep} 公里，發生時間
{eq_time}。', img]
            break
        return msg
    except:
        return msg

# LINE push 訊息函式
def push_message(msg, uid, token):
    headers = {'Authorization':f'Bearer {token}','Content-
Type':'application/json'}
    body = {
    'to':uid,
    'messages':[{
            "type": "text",
            "text": msg
        }]
    }
    req = requests.request('POST', 'https://api.line.me/v2/bot/
message/push', headers=headers,data=json.dumps(body).encode('utf-8'))
    print(req.text)

# LINE 回傳訊息函式
def reply_message(msg, rk, token):
    headers = {'Authorization':f'Bearer {token}','Content-Type':'
application/json'}
    body = {
    'replyToken':rk,
    'messages':[{
            "type": "text",
            "text": msg
```

```
        }]
    }
    req = requests.request('POST', 'https://api.line.me/v2/bot/
message/reply', headers=headers,data=json.dumps(body).encode('utf-8'))
    print(req.text)

# LINE 回傳圖片函式
def reply_image(msg, rk, token):
    headers = {'Authorization':f'Bearer {token}','Content-
Type':'application/json'}
    body = {
    'replyToken':rk,
    'messages':[{
        'type': 'image',
        'originalContentUrl': msg,
        'previewImageUrl': msg
    }]
    }
    req = requests.request('POST', 'https://api.line.me/v2/bot/
message/reply', headers=headers,data=json.dumps(body).encode('utf-8'))
    print(req.text)

# 目前天氣函式
def current_weather(address):
    city_list, area_list, area_list2 = {}, {}, {}
    msg = '找不到氣象資訊。'

    def get_data(url):
        w_data = requests.get(url)
        w_data_json = w_data.json()
        location = w_data_json['cwbopendata']['location']
        for i in location:
            name = i['locationName']
            city = i['parameter'][0]['parameterValue']
            area = i['parameter'][2]['parameterValue']
            temp = check_data(i['weatherElement'][3]['elementValue']
['value'])
            humd = check_data(round(float(i['weatherElement'][4]
['elementValue']['value'] )*100 ,1))
            r24 = check_data(i['weatherElement'][6]['elementValue']
```

```
['value'])
            if area not in area_list:
                area_list[area] = {'temp':temp, 'humd':humd,'r24':r24}
            if city not in city_list:
                city_list[city] = {'temp':[], 'humd':[], 'r24':[]}
            city_list[city]['temp'].append(temp)
            city_list[city]['humd'].append(humd)
            city_list[city]['r24'].append(r24)

    def check_data(e):
        return False if float(e)<0 else float(e)

    def msg_content(loc, msg):
        a = msg
        for i in loc:
            if i in address:
                temp = f"氣溫 {loc[i]['temp']} 度，" if loc[i]['temp']
!= False else ''
                humd = f"相對濕度 {loc[i]['humd']}%，" if loc[i]['humd']
!= False else ''
                r24 = f" 累積雨量 {loc[i]['r24']}mm" if loc[i]['r24'] !=
False else ''
                description = f'{temp}{humd}{r24}'.strip('，')
                a = f'{description}。'
                break
        return a

    try:
        code = ' 你的氣象資料授權碼 '
        get_data(f'https://opendata.cwb.gov.tw/fileapi/v1/opendataapi/
O-A0001-001?Authorization={code}&downloadType=WEB&format=JSON')
        get_data(f'https://opendata.cwb.gov.tw/fileapi/v1/opendataapi/
O-A0003-001?Authorization={code}&downloadType=WEB&format=JSON')

        for i in city_list:
            if i not in area_list2:
                area_list2[i] = {'temp':round(statistics.mean(city_
list[i]['temp']),1),
                                 'humd':round(statistics.mean(city_
list[i]['humd']),1),
```

```
                                      'r24':round(statistics.mean(city_
list[i]['r24']),1)
                            }
        msg = msg_content(area_list2, msg)
        msg = msg_content(area_list, msg)
        return msg
    except:
        return msg


def forecast(address):
    area_list = {}
    json_api = {"宜蘭縣":"F-D0047-001"," 桃園市 ":"F-D0047-005"," 新竹縣
":"F-D0047-009"," 苗栗縣 ":"F-D0047-013",
            " 彰化縣 ":"F-D0047-017"," 南投縣 ":"F-D0047-021"," 雲林縣 ":"F-
D0047-025"," 嘉義縣 ":"F-D0047-029",
            " 屏東縣 ":"F-D0047-033"," 臺東縣 ":"F-D0047-037"," 花蓮縣 ":"F-
D0047-041"," 澎湖縣 ":"F-D0047-045",
            " 基隆市 ":"F-D0047-049"," 新竹市 ":"F-D0047-053"," 嘉義市 ":"F-
D0047-057"," 臺北市 ":"F-D0047-061",
            " 高雄市 ":"F-D0047-065"," 新北市 ":"F-D0047-069"," 臺中市 ":"F-
D0047-073"," 臺南市 ":"F-D0047-077",
            " 連江縣 ":"F-D0047-081"," 金門縣 ":"F-D0047-085"}
    msg = ' 找不到天氣預報資訊。'
    try:
        code = ' 你的氣象開放平台授權碼 '
        url = f'https://opendata.cwb.gov.tw/fileapi/v1/opendataapi/
F-C0032-001?Authorization={code}&downloadType=WEB&format=JSON'
        f_data = requests.get(url)
        f_data_json = f_data.json()
        location = f_data_json['cwbopendata']['dataset']['location']
        for i in location:
            city = i['locationName']
            wx8 = i['weatherElement'][0]['time'][0]['parameter']
['parameterName']
            mint8 = i['weatherElement'][1]['time'][0]['parameter']
['parameterName']
            maxt8 = i['weatherElement'][2]['time'][0]['parameter']
['parameterName']
            ci8 = i['weatherElement'][2]['time'][0]['parameter']
['parameterName']
```

```
                pop8 = i['weatherElement'][2]['time'][0]['parameter']
['parameterName']
            area_list[city] = f' 未來 8 小時 {wx8}，最高溫 {maxt8} 度，最
低溫 {mint8} 度，降雨機率 {pop8} %'
        for i in area_list:
            if i in address:
                msg = area_list[i]
                url = f'https://opendata.cwb.gov.tw/api/v1/rest/datastore/
{json_api[i]}?Authorization={code}&elementName=WeatherDescription'
                f_data = requests.get(url)
                f_data_json = f_data.json()
                location = f_data_json['records']['locations'][0]
['location']
                break
        for i in location:
            city = i['locationName']
            wd = i['weatherElement'][0]['time'][1]['elementValue'][0]
['value']
            if city in address:
                msg = f' 未來八小時天氣 {wd}'
                break
        return msg
    except:
        return msg


# 空氣品質函式
def aqi(address):
    city_list, site_list ={}, {}
    msg = ' 找不到空氣品質資訊。'
    try:
        url = 'https://data.epa.gov.tw/api/v1/aqx_
p_432?limit=1000&api_key=9be7b239-557b-4c10-9775-
78cadfc555e9&sort=ImportDate%20desc&format=json'
        a_data = requests.get(url)          # 使用 get 方法透過空氣品質指標
API 取得內容
        a_data_json = a_data.json()        # json 格式化訊息內容
        for i in a_data_json['records']:# 依序取出 records 內容的每個項目
            city = i['County']             # 取出縣市名稱
            if city not in city_list:
```

```
                    city_list[city]=[]        # 以縣市名稱為 key，準備存入串列資料
            site = i['SiteName']          # 取出鄉鎮區域名稱
            aqi = int(i['AQI'])           # 取得 AQI 數值
            status = i['Status']          # 取得空氣品質狀態
            site_list[site] = {'aqi':aqi, 'status':status}  # 記錄鄉鎮
區域空氣品質
            city_list[city].append(aqi) # 將各個縣市裡的鄉鎮區域空氣 aqi
數值，以串列方式放入縣市名稱的變數裡
        for i in city_list:
            if i in address: # 如果地址裡包含縣市名稱的 key，就直接使用對應
的內容
                # 參考 https://airtw.epa.gov.tw/cht/Information/
Standard/AirQualityIndicator.aspx
                aqi_val = round(statistics.mean(city_list[i]),0)  # 計
算平均數值，如果找不到鄉鎮區域，就使用縣市的平均值
                aqi_status = ''   # 手動判斷對應的空氣品質說明文字
                if aqi_val<=50: aqi_status = '良好'
                elif aqi_val>50 and aqi_val<=100: aqi_status = '普通'
                elif aqi_val>100 and aqi_val<=150: aqi_status = '對敏
感族群不健康'
                elif aqi_val>150 and aqi_val<=200: aqi_status = '對所
有族群不健康'
                elif aqi_val>200 and aqi_val<=300: aqi_status = '非常
不健康'
                else: aqi_status = '危害'
                msg = f'空氣品質{aqi_status} ( AQI {aqi_val} )。' # 定
義回傳的訊息
                break
        for i in site_list:
            if i in address:   # 如果地址裡包含鄉鎮區域名稱的 key，就直接使用
對應的內容
                msg = f'空氣品質{site_list[i]["status"]} ( AQI {site_
list[i]["aqi"]} )。'
                break
        return msg    # 回傳 msg
    except:
        return msg    # 如果取資料有發生錯誤，直接回傳 msg
```

(範例程式碼：ch7/code08.py)

完成後執行程式 (使用 Colab + ngrok 需要重新設定 Webhook)，LINE 裡面傳送地址資訊，除了會回覆目前所在位置的即時氣、天氣預報以及空氣品質的資訊。

7-4 氣象機器人 (4) - 加入圖文選單和部署程式

在「氣象機器人 (1)～氣象機器人 (3)」的三個小節裡，已經完整實出 LINE 的氣象機器人，接下來將會替氣象機器人新增圖文選單，只要點擊圖文選單的按鈕就能讓機器人快速提供氣象資訊，最後再將機器人部署到 Google Cloud Functions 裡，成為可以 24 小時運作不間斷的氣象機器人。

下載範例圖片放入和程式相同的資料夾或 Google 雲端硬碟裡（和 Colab 程式放在同一個目錄）。

> **範例圖片下載：**
>
> https://steam.oxxostudio.tw/download/python/line-bot-weather-demo.jpg

參考「6-3、建立圖文選單」一節，「**依序輸入**」下列的程式碼，替 LINE 氣象機器人加入圖文選單，第一組程式碼，設定圖片與按鈕位置，產生圖文選單 id（輸入自己的 access token）：

```
import requests
import json
headers = {'Authorization':'Bearer 你的 access token','Content-
Type':'application/json'}

body = {
```

```
    'size': {'width': 2500, 'height': 640},      # 設定尺寸
    'selected': 'true',                          # 預設是否顯示
    'name': 'bbb',                               # 選單名稱
    'chatBarText': 'b',                          # 選單在 LINE 顯示的標題
    'areas':[                                    # 選單內容
        {
            'bounds': {'x': 0, 'y': 0, 'width': 1250, 'height': 640},
# 選單位置與大小
            'action': {'type': 'uri', 'uri': 'https://line.me/R/nv/
location/'}   # 點擊後開啟地圖定位，傳送位置資訊
        },
        {
            'bounds': {'x': 1251, 'y': 0, 'width':625, 'height': 640},
# 選單位置與大小
            'action': {'type': 'message', 'text':'雷達回波圖 '}
# 點擊後傳送文字
        },
        {
            'bounds': {'x': 1879, 'y': 0, 'width':625, 'height': 640},
# 選單位置與大小
            'action': {'type': 'message', 'text':'地震資訊 '}
# 點擊後傳送文字
        }
    ]
}
# 向指定網址發送 request
req = requests.request('POST', 'https://api.line.me/v2/bot/richmenu',
                    headers=headers,data=json.dumps(body).
encode('utf-8'))
# 印出得到的結果
print(req.text)
```

(範例程式碼：ch7/code09.py)

第二組程式碼，將圖文選單綁定圖片 (輸入自己的 access token 和圖文選單 id)：

```
from linebot import  LineBotApi, WebhookHandler

line_bot_api = LineBotApi(' 你的 access token')

# import os
# os.chdir('/content/drive/MyDrive/Colab Notebooks') # Colab 換路徑使用

# 開啟對應的圖片
with open('line-bot-weather-demo.jpg', 'rb') as f:
    line_bot_api.set_rich_menu_image(' 你的圖文選單 ID', 'image/jpeg', f)
```
(範例程式碼：ch7/code10.py)

第三組程式碼，將圖文選單與 LINE BOT 綁定 (輸入自己的 access token 和圖文選單 id)：

```
import requests

headers = {"Authorization":"Bearer 你的 access token", "Content-Type":"application/json"}

req = requests.request('POST', 'https://api.line.me/v2/bot/user/all/richmenu/ 圖文選單 id', headers=headers)

print(req.text)
```
(範例程式碼：ch7/code11.py)

完成後，在氣象機器人的聊天畫面下方，就會出現圖文選單，點擊圖文選單，就會出現對應的動作。

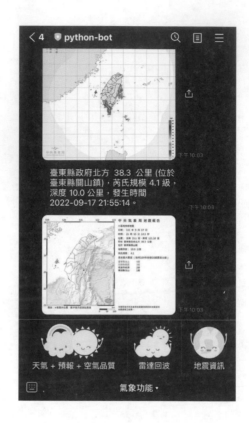

7-5 部署程式到 **Google Cloud Funcions**

參考第十章「使用 Google Cloud Functions」，進入 Google Cloud functions
並開啟一個新的專案，觸發條件勾選允許未經驗證的調用 (注意，使用
Google Cloud Functions 只有第一一年免費，第二年開始基本費用為每個月
0.01 美金)。

環境設定為 Python 3.7 ～ 3.9、進入點為 linebot。

點擊 requirements.txt，加入 line-bot-sdk 和 requests。

```
# Function dependencies, for example:
# package>=version
line-bot-sdk
requests
```

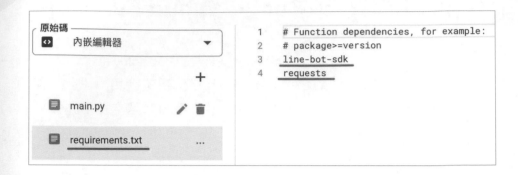

點擊 main.py，將剛剛所撰寫的程式移植到 Cloud Functions 裡，填入自己的 access token、channel secret 和氣象資料平台的授權碼（詳細說明寫在程式碼註解中）。

```python
from linebot import LineBotApi, WebhookHandler
from linebot.models import TextSendMessage, StickerSendMessage,
ImageSendMessage, LocationSendMessage
import requests, statistics, json, time

access_token = '你的 access token'
channel_secret = '你的 channel_secret'

def linebot(request):
    body = request.get_data(as_text=True)
    try:
        json_data = json.loads(body)                      # json 格式化訊息內容
        line_bot_api = LineBotApi(access_token)          # 確認 token 是否正確
        handler = WebhookHandler(channel_secret)
        signature = request.headers['X-Line-Signature']  # 加入回傳的
headers
        handler.handle(body, signature)                   # 綁定訊息回傳
的相關資訊
        reply_token = json_data['events'][0]['replyToken'] # 取得回傳訊
息的 Token ( reply message 使用 )
        user_id = json_data['events'][0]['source']['userId']  # 取得使
用者 ID ( push message 使用 )
```

```
        print(json_data)
        if 'message' in json_data['events'][0]:                    # 如果傳
送的是 message
            if json_data['events'][0]['message']['type'] ==
'location':  # 如果 message 的類型是地圖 location
                address = json_data['events'][0]['message']
['address'].replace('台','臺')    # 取出地址資訊，並將「台」換成「臺」
                reply_message(f'{address}\n\n{aqi(address)}\n\
n{current_weather(address)}\n\n{forecast(address)}', reply_token,
access_token) # 回覆爬取到的相關氣象資訊
            if json_data['events'][0]['message']['type'] == 'text':
# 如果 message 的類型是文字 text
                text = json_data['events'][0]['message']['text']
# 取出文字
                if text == '地震資訊' or text == '地震':
# 如果是地震相關的文字
                    msg = earth_quake()
# 爬取地震資訊
                    push_message(msg[0], user_id, access_token)
# 傳送地震資訊（用 push 方法，因為 reply 只能用一次）
                    reply_image(msg[1], reply_token, access_token)
# 傳送地震圖片（用 reply 方法）
                elif text == '雷達回波圖' or text == '雷達回波':
# 如果是雷達回波圖相關的文字
                    # 傳送雷達回波圖（加上時間戳記）
                    reply_image(f'https://cwbopendata.s3.ap-northeast-1.
amazonaws.com/MSC/O-A0058-001.png?{time.time_ns()}',
reply_token, access_token)
                else:
                    reply_message(text, reply_token, access_token)
# 如果是一般文字，直接回覆同樣的文字
    except:
        print('error')
    return 'OK'


# 氣象預報函式
def forecast(address):
    area_list = {}
    # 將主要縣市個別的 JSON 代碼列出
```

```
    json_api = {" 宜蘭縣 ":"F-D0047-001"," 桃園市 ":"F-D0047-005"," 新竹縣
":"F-D0047-009"," 苗栗縣 ":"F-D0047-013",
            " 彰化縣 ":"F-D0047-017"," 南投縣 ":"F-D0047-021"," 雲林縣 ":"F-
D0047-025"," 嘉義縣 ":"F-D0047-029",
            " 屏東縣 ":"F-D0047-033"," 臺東縣 ":"F-D0047-037"," 花蓮縣 ":"F-
D0047-041"," 澎湖縣 ":"F-D0047-045",
            " 基隆市 ":"F-D0047-049"," 新竹市 ":"F-D0047-053"," 嘉義市 ":"F-
D0047-057"," 臺北市 ":"F-D0047-061",
            " 高雄市 ":"F-D0047-065"," 新北市 ":"F-D0047-069"," 臺中市 ":"F-
D0047-073"," 臺南市 ":"F-D0047-077",
            " 連江縣 ":"F-D0047-081"," 金門縣 ":"F-D0047-085"}
    msg = ' 找不到天氣預報資訊。'      # 預設回傳訊息
    try:
        code = ' 你的氣象資料授權碼 '
        url = f'https://opendata.cwb.gov.tw/fileapi/v1/opendataapi/{cod
e}&downloadType=WEB&format=JSON'
        f_data = requests.get(url)     # 取得主要縣市預報資料
        f_data_json = f_data.json()  # json 格式化訊息內容
        location = f_data_json['cwbopendata']['dataset']['location']
# 取得縣市的預報內容
        for i in location:
            city = i['locationName']     # 縣市名稱
            wx8 = i['weatherElement'][0]['time'][0]['parameter']
['parameterName']     # 天氣現象
            mint8 = i['weatherElement'][1]['time'][0]['parameter']
['parameterName']   # 最低溫
            maxt8 = i['weatherElement'][2]['time'][0]['parameter']
['parameterName']   # 最高溫
            ci8 = i['weatherElement'][2]['time'][0]['parameter']
['parameterName']     # 舒適度
            pop8 = i['weatherElement'][2]['time'][0]['parameter']
['parameterName']    # 降雨機率
            area_list[city] = f' 未來 8 小時 {wx8}，最高溫 {maxt8} 度，最
低溫 {mint8} 度，降雨機率 {pop8} %'  # 組合成回傳的訊息，存在以縣市名稱為 key
的字典檔裡
        for i in area_list:
            if i in address:            # 如果使用者的地址包含縣市名稱
                msg = area_list[i]  # 將 msg 換成對應的預報資訊
                # 將進一步的預報網址換成對應的預報網址
```

```
            url = f'https://opendata.cwb.gov.tw/api/v1/rest/
datastore/{json_api[i]}?Authorization=CWB-DCA7061B-EEEC-496E-9B46-DC75
49FA5F88&elementName=WeatherDescription'
            f_data = requests.get(url)  # 取得主要縣市裡各個區域鄉鎮的
氣象預報
            f_data_json = f_data.json() # json 格式化訊息內容
            location = f_data_json['records']['locations'][0]
['location']    # 取得預報內容
            break
        for i in location:
            city = i['locationName']   # 取得縣市名稱
            wd = i['weatherElement'][0]['time'][1]['elementValue'][0]
['value']   # 綜合描述
            if city in address:        # 如果使用者的地址包含鄉鎮區域名稱
                msg = f' 未來八小時天氣 {wd}' # 將 msg 換成對應的預報資訊
                break
        return msg  # 回傳 msg
    except:
        return msg  # 如果取資料有發生錯誤，直接回傳 msg


# 目前天氣函式
def current_weather(address):
    city_list, area_list, area_list2 = {}, {}, {} # 定義好待會要用的變數
    msg = ' 找不到氣象資訊。'                        # 預設回傳訊息

    # 定義取得資料的函式
    def get_data(url):
        w_data = requests.get(url)    # 爬取目前天氣網址的資料
        w_data_json = w_data.json()   # json 格式化訊息內容
        location = w_data_json['cwbopendata']['location']   # 取出對應地
點的內容
        for i in location:
            name = i['locationName']                        # 測站地點
            city = i['parameter'][0]['parameterValue']   # 城市
            area = i['parameter'][2]['parameterValue']   # 行政區
            temp = check_data(i['weatherElement'][3]['elementValue']
['value'])                    # 氣溫
            humd = check_data(round(float(i['weatherElement'][4]
['elementValue']['value'] )*100 ,1)) # 相對濕度
```

7-59

```
            r24 = check_data(i['weatherElement'][6]['elementValue']
['value'])                              # 累積雨量
            if area not in area_list:
                area_list[area] = {'temp':temp, 'humd':humd,
'r24':r24}   # 以鄉鎮區域為 key，儲存需要的資訊
            if city not in city_list:
                city_list[city] = {'temp':[], 'humd':[], 'r24':[]}
# 以主要縣市名稱為 key，準備紀錄裡面所有鄉鎮的數值
            city_list[city]['temp'].append(temp)    # 記錄主要縣市裡鄉鎮區
域的溫度（串列格式）
            city_list[city]['humd'].append(humd)    # 記錄主要縣市裡鄉鎮區
域的濕度（串列格式）
            city_list[city]['r24'].append(r24)      # 記錄主要縣市裡鄉鎮區
域的雨量（串列格式）

    # 定義如果數值小於 0，回傳 False 的函式
    def check_data(e):
        return False if float(e)<0 else float(e)

    # 定義產生回傳訊息的函式
    def msg_content(loc, msg):
        a = msg
        for i in loc:
            if i in address: # 如果地址裡存在 key 的名稱
                temp = f"氣溫 {loc[i]['temp']} 度，" if loc[i]['temp']
!= False else ''
                humd = f"相對濕度 {loc[i]['humd']}%，" if loc[i]['humd']
!= False else ''
                r24 = f"累積雨量 {loc[i]['r24']}mm" if loc[i]['r24'] !=
False else ''
                description = f'{temp}{humd}{r24}'.strip('，')
                a = f'{description}。' # 取出 key 的內容作為回傳訊息使用
                break
        return a

    try:
        # 因為目前天氣有兩組網址，兩組都爬取
        code = '你的氣象資料授權碼，'
        get_data(f'https://opendata.cwb.gov.tw/fileapi/v1/opendataapi/
```

```
O-A0001-001?Authorization={code}&downloadType=WEB&format=JSON')
        get_data(f'https://opendata.cwb.gov.tw/fileapi/v1/opendataapi/
O-A0003-001?Authorization={code}&downloadType=WEB&format=JSON')


        for i in city_list:
            if i not in area_list2:  # 將主要縣市裡的數值平均後，以主要縣市
名稱為 key，再度儲存一次，如果找不到鄉鎮區域，就使用平均數值
                area_list2[i] = {'temp':round(statistics.mean(city_
list[i]['temp']),1),
                                 'humd':round(statistics.mean(city_
list[i]['humd']),1),
                                 'r24':round(statistics.mean(city_
list[i]['r24']),1)
                                }
        msg = msg_content(area_list2, msg)   # 將訊息改為「大縣市」
        msg = msg_content(area_list, msg)    # 將訊息改為「鄉鎮區域」
        return msg     # 回傳 msg
    except:
        return msg        # 如果取資料有發生錯誤，直接回傳 msg


# 空氣品質函式
def aqi(address):
    city_list, site_list ={}, {}
    msg = '找不到空氣品質資訊。'
    try:
        url = 'https://data.epa.gov.tw/api/v1/aqx_
p_432?limit=1000&api_key=9be7b239-557b-4c10-9775-
78cadfc555e9&sort=ImportDate%20desc&format=json'
        a_data = requests.get(url)              # 使用 get 方法透過空氣品
質指標 API 取得內容
        a_data_json = a_data.json()             # json 格式化訊息內容
        for i in a_data_json['records']:        # 依序取出 records 內容的
每個項目
            city = i['County']                  # 取出縣市名稱
            if city not in city_list:
                city_list[city]=[]              # 以縣市名稱為 key，準備存
入串列資料
            site = i['SiteName']                # 取出鄉鎮區域名稱
            aqi = int(i['AQI'])                 # 取得 AQI 數值
```

```
            status = i['Status']                    # 取得空氣品質狀態
            site_list[site] = {'aqi':aqi, 'status':status}   # 記錄鄉鎮
區域空氣品質

            city_list[city].append(aqi)                   # 將各個縣市裡的鄉鎮區域空
氣 aqi 數值,以串列方式放入縣市名稱的變數裡
        for i in city_list:
            if i in address:  # 如果地址裡包含縣市名稱的 key,就直接使用對應
的內容
                # https://airtw.epa.gov.tw/cht/Information/Standard/
AirQualityIndicator.aspx
                aqi_val = round(statistics.mean(city_list[i]),0)   # 計
算平均數值,如果找不到鄉鎮區域,就使用縣市的平均值
                aqi_status = ''    # 手動判斷對應的空氣品質說明文字
                if aqi_val<=50: aqi_status = '良好'
                elif aqi_val>50 and aqi_val<=100: aqi_status = '普通'
                elif aqi_val>100 and aqi_val<=150: aqi_status = '對敏
感族群不健康'
                elif aqi_val>150 and aqi_val<=200: aqi_status = '對所
有族群不健康'
                elif aqi_val>200 and aqi_val<=300: aqi_status = '非常
不健康'
                else: aqi_status = '危害'
                msg = f'空氣品質 {aqi_status} ( AQI {aqi_val} )。' # 定
義回傳的訊息
                break
        for i in site_list:
            if i in address:   # 如果地址裡包含鄉鎮區域名稱的 key,就直接使用
對應的內容
                msg = f'空氣品質 {site_list[i]["status"]} ( AQI {site_
list[i]["aqi"]} )。'
                break
        return msg     # 回傳 msg
    except:
        return msg     # 如果取資料有發生錯誤,直接回傳 msg

# 地震資訊函式
def earth_quake():
    msg = ['找不到地震資訊','https://example.com/demo.jpg']
    try:
```

```
        code = '你的氣象資料授權碼'
        url = f'https://opendata.cwb.gov.tw/api/v1/rest/datastore/
E-A0016-001?Authorization={code}'
        e_data = requests.get(url)        # 爬取地震資訊網址
        e_data_json = e_data.json()       # json 格式化訊息內容
        eq = e_data_json['records']['earthquake']  # 取出地震資訊
        for i in eq:
            loc = i['earthquakeInfo']['epiCenter']['location']
# 地震地點
            val = i['earthquakeInfo']['magnitude']['magnitudeValue']
# 地震規模
            dep = i['earthquakeInfo']['depth']['value']
# 地震深度
            eq_time = i['earthquakeInfo']['originTime']
# 地震時間
            img = i['reportImageURI']
# 地震圖
            msg = [f'{loc}，芮氏規模 {val} 級，深度 {dep} 公里，發生時間 {eq_
time}。', img]
            break       # 取出第一筆資料後就 break
        return msg      # 回傳 msg
    except:
        return msg      # 如果取資料有發生錯誤，直接回傳 msg


# LINE 回傳訊息函式
def reply_message(msg, rk, token):
    headers = {'Authorization':f'Bearer {token}','Content-
Type':'application/json'}
    body = {
    'replyToken':rk,
    'messages':[{
            "type": "text",
            "text": msg
        }]
    }
    req = requests.request('POST', 'https://api.line.me/v2/bot/
message/reply', headers=headers,data=json.dumps(body).encode('utf-8'))
    print(req.text)
```

```python
# LINE 回傳圖片函式
def reply_image(msg, rk, token):
    headers = {'Authorization':f'Bearer {token}','Content-
Type':'application/json'}
    body = {
    'replyToken':rk,
    'messages':[{
        'type': 'image',
        'originalContentUrl': msg,
        'previewImageUrl': msg
    }]
    }
    req = requests.request('POST', 'https://api.line.me/v2/bot/
message/reply', headers=headers,data=json.dumps(body).encode('utf-8'))
    print(req.text)

# LINE push 訊息函式
def push_message(msg, uid, token):
    headers = {'Authorization':f'Bearer {token}','Content-
Type':'application/json'}
    body = {
    'to':uid,
    'messages':[{
        "type": "text",
        "text": msg
    }]
    }
    req = requests.request('POST', 'https://api.line.me/v2/bot/
message/push', headers=headers,data=json.dumps(body).encode('utf-8'))
    print(req.text)

# LINE push 圖片函式
def push_image(msg, uid, token):
    headers = {'Authorization':f'Bearer {token}','Content-
Type':'application/json'}
    body = {
    'to':uid,
    'messages':[{
        'type': 'image',
```

```
        'originalContentUrl': msg,
        'previewImageUrl': msg
    }]
  }
  req = requests.request('POST', 'https://api.line.me/v2/bot/
message/push', headers=headers,data=json.dumps(body).encode('utf-8'))
  print(req.text)
```
(範例程式碼：ch7/code12.py)

```
原始碼
    內嵌編輯器           ▼
                    +
  main.py           ✎ 🗑
  requirements.txt  ...
```
```
1   from linebot import LineBotApi, WebhookHandler
2   from linebot.models import TextSendMessage, StickerSendMessage, ImageSendMessage, LocationSendMessage
3   import requests, statistics, json, time
4
5   access_token = '你的 access token'
6   channel_secret = '你的 channel_secret'
7
8   def linebot():
9       #你的 linebot 函式內容
10
11  # 氣象預報函式
12  def forecast(address):
13
14  # 目前天氣函式
15  def current_weather(address):
16
17  # 空氣品質函式
18  def aqi(address):
19
20  # 地震資訊函式
```

完成後點擊部署，部署完成後前方會出現綠色打勾圖示。

```
(···)  Cloud Functions        函式        ➕建立函式    ↻重新整理

≡ 篩選   篩選函式
```

	●	環境	名稱 ↑	區域	觸發條件	執行階段
☐	✓	1st gen	**py-test**	asia-east1	HTTP	Python 3.7

點擊進入專案，點選「觸發條件」，複製觸發網址，回到 LINE Developer 控制台，將 Webhook 更新為觸發的網址，驗證通過後，就完成了一個可以 24 小時運作的 LINE 氣象機器人。

 小結

透過這個章節的教學，就能打造出一個具有圖文選單，且能 24 小時運作的 LINE 氣象機器人，如果活用相關的技巧，就能讓機器人的功能更為強大囉！

8

串接 Dialogflow
打造聊天機器人

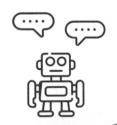

前 言

這篇教學會介紹如何使用 Dialogflow，透過 Dialogflow 串接 Python 伺服器，最後搭配 LINE BOT 的程式運作，實作出一個能夠理解自然語意的聊天機器人。

本章節的範例程式碼：

https://github.com/oxxostudio/book-code/tree/master/linebot/ch8

兩點注意事項：

要實作 LINE 的聊天機器人需要先註冊 LINE 開發者帳號 (取得 Access Token 和 Channel secret)，以及架設 Webhook 與 LINE 串接，請先閱讀本書第二章「建立 LINE BOT」和第三章「開發環境設定＆串接 LINE BOT」，完成相關步驟後再進行閱讀。

實作過程中會需要使用一些 Python 基本語法，例如串列與字典解析、字串格式化、邏輯、迴圈 ... 等，如果對於語法不熟悉，請前往 https://steam.oxxostudio. tw/category/python/ 進行相關基礎語法學習。

8-1 使用 **Google Dialogflow**

Dialogflow 是一個 Google 的開發工具，主要作用是進行自然語言處理的服務，能在不需撰寫程式的狀況下，透過 Dialogflow 快速打造聊天機器人，這個章節會介紹如何使用 Dialogflow，並在 Dialogflow 裡建立語意的資料庫。

認識 Dialogflow

Dialogflow 的前身是 Speaktoit 的 Api.ai，是一個 Google 的開發工具，**在 Dialogflow 裡可以加入許多的對話「意圖 Intent」，每個意圖可以包涵許多不同的語句**，例如「今天天氣好嗎？」和「今天天氣如何？」是屬於「問天氣」的對話意圖，透過許多的語句，就能進行自然語言的處理，就算語句資料庫中沒有「天氣怎麼樣」這句話，輸入時仍然會將其歸類到「問天氣」的對話意圖裡，**當建立了足夠的語句和意圖，機器人就很容易理解人類所講的「自然語言」。**

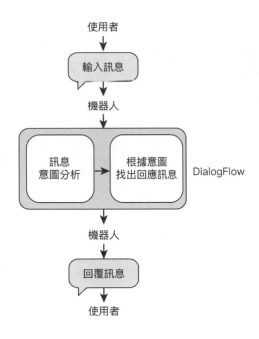

建立意圖資料庫之後，**Dialogflow 提供許多聊天機器人的串接方式，例如 Facebook、Slack、LINE... 等**，只要簡單幾個步驟，就能在各大平台上創建聊天機器人，此外，Dialogflow 也支援串接 Webhook，可以讓使用者在聊天時串接自己的服務，透過自己的服務進行更多後端的應用，例如爬蟲、分析 ... 等。

Dialogflow 提供「基本免費」的使用（ES Agent Trial Edition 版本），但如果請求數（request）數過多，或需要額外串接 Google Cloud 相關服務，就必須要負擔額外的費用（費用參考：https://cloud.google.com/dialogflow/pricing?hl=zh-tw）。

開始使用 Dialogflow

前往 Dialogflow 平台，使用自己的 Google 帳號登入（第一次使用需要先同意條款）。

> Dialogflow：https://dialogflow.cloud.google.com/

進入後，點擊「Creat Agent」就可以建立第一個聊天代理人 Agent。

建立 Agent 時需要輸入名稱、設定語系（如果聊天機器人的主要自然語言為中文，就選擇中文語系）、設定時區（如果在台灣就設定為香港時區）。

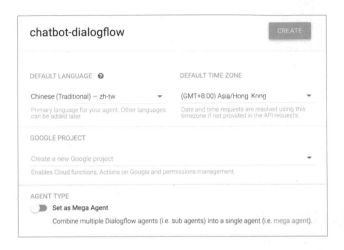

完成後如果出現 Intents 的頁籤內容，左側也出現各種選單，表示 Agent 已經建立完成。

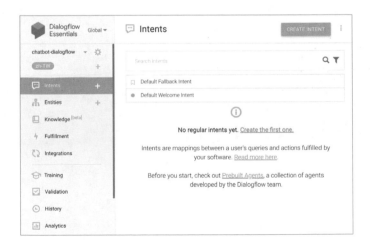

建立對話意圖 (Intent)

點擊 Intents 頁籤,從中可以建立「對話意圖」,**對話意圖 Intent 的意思是「某一段話代表什麼意思」**,例如「早安」、「大家早」、「**Good morning**」這三句話都可以看做「說早安」的對話意圖,在 Dialogflow 裡預設有 Defult Fallback Intent 和 Default Welcome Intent 兩組對話意圖。

Defult Fallback Intent 表示「未知的意圖」,也就是如果無法解析傳送訊息的意圖,就會歸類到這一類,這時 Agent 就會從下方所列的訊息裡,自動選擇一個進行回覆 (在 Responses 區塊按下 + 號就可以增加回覆的訊息)。

Default Welcome Intent 表示「**歡迎意圖、打招呼意圖**」，也就是如果輸入了「嗨」、「哈囉」之類的打招呼語句，Agent 就會從下方所列的訊息裡，自動選擇一個進行回覆 (在 Responses 區塊按下 + 號就可以增加回覆的訊息)。

Default Welcome Intent 表示「歡迎意圖、打招呼意圖」，也就是如果輸入了「嗨」、「哈囉」之類的打招呼語句，Agent 就會從下方所列的訊息裡，自動選擇一個進行回覆 (在 Responses 區塊按下 + 號就可以增加回覆的訊息)。

了解原理後,就可以嘗試建立一個名為 Weather 的「問氣象意圖 Intent」,內容只要輸入的尋問氣象相關的語句,就會回答簡單的對應訊息。

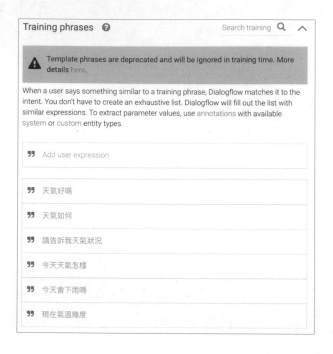

🖐 機器人聊天測試

對話意圖完成後，**從右側上方的 Try it now，就可以輸入一些詢問的語句，輸入後就會看見機器人自動回覆**，如果有發現一些語句不符合意圖，就可以返回相關的 Intent 進行修改。

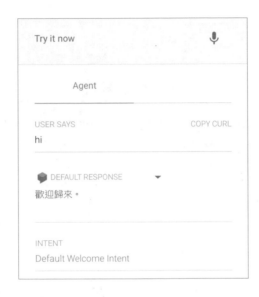

8-2　Dialogflow 串接 Webhook

如果可以在 DialogFlow 裡和聊天機器人聊天，接下來將會介紹如何使用
Python 架設簡單的伺服器 (本機環境、Colab 或 Google Cloud Functions)，
建立 Webhook 網址，並將 DialogFlow 串接 Webhook，透過伺服器做到更
多 DialogFlow 做不到的事情。

Dialogflow 與 WebHook 的關係

當使用者與串接 Dialogflow 的機器人聊天時，如果 Dialogflow 有串接
Webhook，則會發生下列的步驟：

Step 1：向串接 Dialogflow 的機器人發送訊息。

Step 2：機器人收到訊息後，將訊息透過 Dialogflow 解析語意。

Step 3：解析語意後，將解析的語意透過 Webhook 傳送到使用者 Python 的伺
　　　　服器，根據自訂義的邏輯處理語意內容。

Step 4：處理語意內容後，將結果再透過 Webhook 回傳到 Dialogflow。

Step 5：Dialogflow 收到結果後，透過串接的機器人，將結果的訊息傳送給使用
　　　　者。

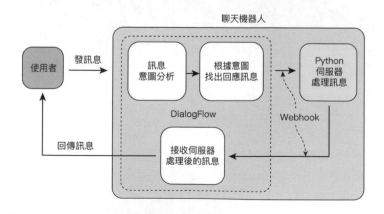

建立 Webhook (本機環境)

參考本書第三章「註冊 ngrok 服務」先行註冊 ngrok 並取得 Token，接著在本機環境使用命令提示字元啟用 ngrok (如果要使用 Colab 請直接往後閱讀 Colab 建立 Webhook)。

```
ngrok http 5000
```

啟用後就會看見 ngrok 服務對應到本機伺服器產生的臨時網址。

```
ngrok by @inconshreveable                                    (Ctrl+C to quit)

Session Status          online
Account                 ohha12345 (Plan: Free)
Update                  update available (version 2.3.40, Ctrl-U to update
Version                 2.3.35
Region                  United States (us)
Web Interface           http://127.0.0.1:4040
Forwarding              http://96be-220-133-228-250.ngrok.io -> http://loc
Forwarding              https://96be-220-133-228-250.ngrok.io -> http://lo

Connections             ttl       opn       rt1       rt5       p50       p90
                        4         0         0.02      0.01      0.01      0.02
```

建立一個新的 Python 檔案，使用下方的程式碼，執行後就能建立一個簡單的伺服器 (如果使用全新的 Python 虛擬環境，需要安裝 Flask)，**當中 / webhook 的入口就是 Webhook 的網址，webhook() 函式的內容會將 Dialogflow 傳送過來的字串轉換成 dict 格式**，取出當中要回應的字串，並在字串後方加上 (webhook) 文字，證明這是透過 Webhook 伺服器處理後的文字訊息。

```
from flask import Flask, request

app = Flask(__name__)

@app.route("/")
```

```python
def home():
    return "<h1>hello world</h1>"

@app.route('/webhook', methods=['POST'])
def webhook():
    # 轉換成 dict 格式
    req = request.get_json()
    print(req)
    # 取得回覆文字
    reText = req['queryResult']['fulfillmentText']
    print(reText)
    # 在回覆的文字後方加上 ( webhook ) 識別
    return {
        "fulfillmentText": f'{reText} ( webhook )',
        "source": "webhookdata"
    }

app.run()
```
(範例程式碼：ch8/code01.py)

程式執行後，開啟瀏覽器，輸入剛剛 ngrok 產生的網址（後方不要加上 /
webhook），如果出現 hello world 的文字，表示順利建立成功，**成功後直接
前往後繼續閱讀「Dialogflow 串接 Webhook」一節**。

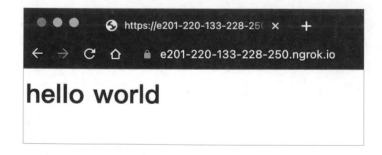

建立 Webhook (Google Colab)

參考本書第三章「註冊 ngrok 服務」，先行註冊 ngrok 並取得 Token，參考
本書第三章「「建立 Webhook (Google Colab)」」，在 Colab 裡安裝並啟
用 ngrok (如果要使用本機環境請直接往前閱讀本機環境建立 Webhook)。

安裝步驟都完成後，在 Colab 裡輸入下方的程式碼，**當中 /webhook 的入
口就是 Webhook 的網址，webhook() 函式的內容會將 Dialogflow 傳送
過來的字串轉換成 dict 格式**，取出當中要回應的字串，並在字串後方加上 (
webhook) 文字，證明這是透過 Webhook 伺服器處理後的文字訊息。

```python
from flask import Flask, request
from flask_ngrok import run_with_ngrok


app = Flask(__name__)
run_with_ngrok(app)        # 連結 ngrok

@app.route("/")
def home():
    return "<h1>hello world</h1>"

@app.route('/webhook', methods=['POST'])
def webhook():
    # 轉換成 dict 格式
    req = request.get_json()
    print(req)
    # 取得回覆文字
    reText = req['queryResult']['fulfillmentText']
    print(reText)
    # 在回覆的文字後方加上 ( webhook ) 識別
    return {
        "fulfillmentText": f'{reText} ( webhook )',
        "source": "webhookdata"
    }


app.run()
```
(範例程式碼：ch8/code02.py)

點擊 Colab 的執行按鈕，就會得到一串 ngrok 對應的網址，這串網址就是要與 Dialogflow 串接的 Webhook，表示順利建立成功，**成功後直接前往後繼續閱讀「Dialogflow 串接 Webhook」一節**。

```
* Serving Flask app "__main__" (lazy loading)
* Environment: production
  WARNING: This is a development server. Do not use it in a production deployment.
  Use a production WSGI server instead.
* Debug mode: off
INFO:werkzeug: * Running on http://127.0.0.1:5000/ (Press CTRL+C to quit)
 * Running on http://2576-34-125-114-163.ngrok.io
 * Traffic stats available on http://127.0.0.1:4040
INFO:werkzeug:127.0.0.1 - - [01/Sep/2022 04:44:16] "GET / HTTP/1.1" 200 -
INFO:werkzeug:127.0.0.1 - - [01/Sep/2022 04:44:16] "GET /favicon.ico HTTP/1.1" 404 -
```

建立 Webhook (Cloud Functions)

由於使用 Colab + ngrok 所建置的 Webhook，會受限於 Colab 只能運行幾個小時，以及 ngrok 在每次部署都會改變網址的特性，所以無法當作正式的 LINE BOT Webhook (Colab 閒置超過一段時間後還會停止執行並清除安裝的函式庫，需要再次重新安裝，本機環境也會受限於電腦關機或 IP 變動等問題)。

如果要建立一個可以 24 小時不斷運作的 Webhook，就可以選擇 Google Cloud Functions 作為 Python 運作的後台，參考本書第十章「使用 Google Cloud Functions」，新增並啟用一個 Cloud Functions 程式編輯環境，基本設定如下圖所示：

進入編輯畫面後，環境執行階段選擇 Python（選擇 3.9），進入點改成 webhook（可自訂名稱，之後的程式碼裡也要使用同樣的名稱）。

接著輸入下方的程式碼，完成後點擊下方的「部署」，就會將程式部署到 Cloud Functions 裡。

```python
def webhook(request):
    try:
        req = request.get_json()
        reText = req['queryResult']['fulfillmentText']
        return {
            "fulfillmentText": f'{reText} ( webhook )',
            "source": "webhookdata"
        }
    except:
        print(request.args)
```
（範例程式碼：ch8/code03.py）

如果部署順利完成，就會看見該專案前方出現一個綠色打勾圖示，這時切換到「觸發條件」頁籤，就可以看到所需要的 Webhook 網址，**成功後直接前往後繼續閱讀「Dialogflow 串接 Webhook」一節。**

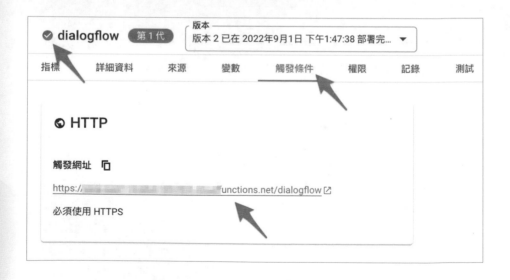

Dialogflow 串接 Webhook

回到 Dialogflow 的專案，進入「Intents」頁籤，點擊需要串接 Webhook 的 Intent，**進入後在最下方勾選「Enable webhook call for this intent」，表示該 Intent 會透過 Webhook 處理後再進行回覆**（如果 Webhook 失敗則會直接套用內容回覆）。

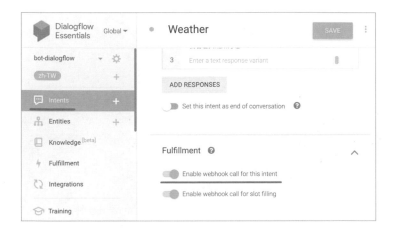

進入「Fulfillment」頁籤，勾選啟用 Webhook，將剛剛產生的 Webhook 貼上並儲存（**如果是本地端或 Colab 網址，後方要加上 /webhook**）

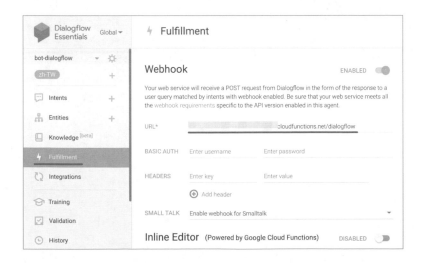

完成後在右側的聊天測試裡，輸入一些文字就可以看見機器人的自動回覆，如果回覆文字的後方有加上 (webhook) 文字，表示已經順利串接 Webhook (額外加上的文字是在 Webhook 伺服器端加入的，可以自行修改程式)。

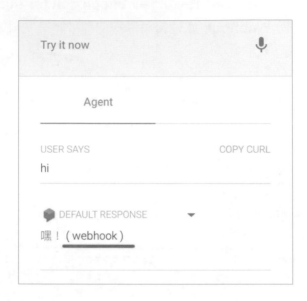

進入 Webhook 伺服器後台，也能看見傳遞的訊息出現在記錄檔裡。

```
INFO:werkzeug:127.0.0.1 - - [01/Sep/2022 06:09:40] "POST /webhook HTTP/1.1" 200 -
{'responseId': '3efe0306-20ff-44af-a6e5-8eca79a38aca-0cab8cdb', 'queryResult': {'queryText': 'hello'
歡迎歸來。
INFO:werkzeug:127.0.0.1 - - [01/Sep/2022 06:09:45] "POST /webhook HTTP/1.1" 200 -
{'responseId': '4c799216-4c9a-4be1-b0a2-682ea4f7c4ad-0cab8cdb', 'queryResult': {'queryText': '天氣',
請你再說一遍。
INFO:werkzeug:127.0.0.1 - - [01/Sep/2022 06:09:50] "POST /webhook HTTP/1.1" 200 -
{'responseId': '8286b7cf-8eb9-45e3-b9c2-697a3b23c5e1-0cab8cdb', 'queryResult': {'queryText': '你好',
嘿!
```

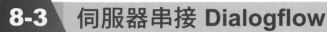

8-3　伺服器串接 Dialogflow

在前面兩個小節裡，已經學會使用 Dialogflow 建立聊天機器人，接下來將會介紹如何使用 Google Cloud 建立金鑰，讓自己的 Python 伺服器，可以透過 API 串接 Dialogflow。

建立並下載金鑰 json

建立 Dialogflow 專案後，同時也會在 Google Cloud Platform 裡建立一個專案，前往 Google Cloud PlatForm 並進入該專案。

前往 Google Cloud Platform 控制台：https://console.cloud.google.com/

點選左上角圖示開啟選單，選擇「IAM 與管理」裡的「服務帳戶」。

如果已經有使用 Dialogflow，會有出現預設的一些服務帳戶，點擊 Dialogflow Integrations 的服務帳戶後方的圖示，選擇「管理金鑰」。

進入後新增金鑰，選擇「建立新的金鑰」。

建立金鑰時選擇 json 檔案，將其下載存放到和 python 伺服器執行的檔案同樣的目錄 (這樣就不用額外處理檔案路徑)

已將私密金鑰儲存至您的電腦中

⚠ 「██████████.json」可用來存取您的雲端資源,因此請妥善存放。瞭解更多最佳做法

關閉

存檔後可以使用編輯器打開查看金鑰 json,內容是 token 之類的資訊 (類似下方格式)。

```
{
  "type": "service_account",
  "project_id": "XXX",
  "private_key_id": "XXX",
  "private_key": "XXXXXXXX",
  "client_email": "XXXXX@appspot.gserviceaccount.com",
  "client_id": "XXXXXXX",
  "auth_uri": "https://accounts.google.com/o/oauth2/auth",
  "token_uri": "https://oauth2.googleapis.com/token",
  "auth_provider_x509_cert_url": "https://www.googleapis.com/oauth2/
v1/certs",
  "client_x509_cert_url": "https://www.googleapis.com/robot/v1/
metadata/x509/XXXXX%40appspot.gserviceaccount.com"
}
```

🖱 串接 Dialogflow (本機環境)

要使用伺服器串接 Dialogflow,必須要先安裝 Google Cloud 相關的函式庫,輸入指令安裝 google-cloud-dialogflow (如果是 Anaconda Jupyter 請使用 !pip 安裝)。

> 因 Google Dialogflow 函式庫無法運行在 Python 3.7 的環境，所以如果遇到無法安裝的情形，請先將 Python 升級為 3.9 以上版本，同理，因為 Colab 預設 Python 3.7，也就無法正確安裝和執行 Google Dialogflow 函式庫。

```
pip install google-cloud-dialogflow
```

執行下方的程式碼 (全新的 Python 虛擬環境需要安裝 Flask)，就能將本機伺服器，串接 Dialogflow，詳細說明寫在程式碼的註解中。

```python
import os
import google.cloud.dialogflow_v2 as dialogflow
from flask import Flask, request

# 讀取下載的金鑰 json
os.environ["GOOGLE_APPLICATION_CREDENTIALS"] = 'dialogflow_key.json'
project_id = 'XXXX'          # dialogflow 的 project id
language = 'zh-TW'           # 語系
session_id = 'oxxostudio'    # 自訂義的 session id

# 建立連接 Dialogflow 的函式
def dialogflowFn(text):
    # 使用 Token 和 dialogflow 建立連線
    session_client = dialogflow.SessionsClient()
    # 連接對應專案
    session = session_client.session_path(project_id, session_id)
    # 設定語系
    text_input = dialogflow.types.TextInput(text=text, language_code=language)
    # 根據語系取得輸入內容
    query_input = dialogflow.types.QueryInput(text=text_input)
    try:
        # 連線 Dialogflow 取得回應資料
        response = session_client.detect_intent(session=session, query_input=query_input)
        # 印出相關資訊
        print("input:", response.query_result.query_text)
```

```python
        print("intent:", response.query_result.intent.display_name)
        print("reply:", response.query_result.fulfillment_text)
        # 回傳回應的文字
        return response.query_result.fulfillment_text
    except:
        return 'error'


app = Flask(__name__)


@app.route("/")
def home():
    # 取得輸入的文字
    text = request.args.get('text')
    # 透過 Dialogflow 得到回應的文字
    reply = dialogflowFn(text)
    return reply


app.run()
```

(範例程式碼：ch8/code04.py)

程式執行後，打開瀏覽器，在網址列輸入伺服器產生的網址，後方加上輸入的文字參數，執行後就可以看見透過 Dialogflow 的回應訊息。

串接 Dialogflow (Cloud Functions)

參考本書第十章「使用 Google Cloud Functions」，新增並啟用一個 Cloud Functions 程式編輯環境，基本設定如下圖所示：

進入編輯畫面後，環境執行階段選擇 Python（選擇 3.9），進入點改成 webhook（可自訂名稱，之後的程式碼裡也要使用同樣的名稱）。

在左側點擊 + 號，新增一個 .json 的檔案（檔名自訂），內容就是剛剛下載的金鑰 json 檔案內容。

點擊 requirement.txt，新增 google-cloud-dialogflow 函式庫。

使用下方的程式碼，完成後點擊「部署」，將程式部署到 Cloud Functions 裡。

```
import os
import google.cloud.dialogflow_v2 as dialogflow

# 金鑰 json
```

```python
os.environ["GOOGLE_APPLICATION_CREDENTIALS"] = 'dialogflow_key.json'
project_id = 'XXXX'        # dialogflow 的 project id
language = 'zh-TW'         # 語系
session_id = 'oxxostudio'  # 自訂義的 session id

def dialogflowFn(text):
    # 使用 Token 和 dialogflow 建立連線
    session_client = dialogflow.SessionsClient()
    # 連接對應專案
    session = session_client.session_path(project_id, session_id)
    # 設定語系
    text_input = dialogflow.types.TextInput(text=text, language_
code=language)
    # 根據語系取得輸入內容
    query_input = dialogflow.types.QueryInput(text=text_input)
    try:
        # 連線 Dialogflow 取得回應資料
        response = session_client.detect_intent(session=session,
query_input=query_input)
        print("input:", response.query_result.query_text)
        print("intent:", response.query_result.intent.display_name)
        print("reply:", response.query_result.fulfillment_text)
        # 回傳回應的文字
        return response.query_result.fulfillment_text
    except:
        return 'error'

def webhook(request):
    try:
        text = request.args.get('text')
        return dialogflowFn(text)
    except:
        print(request.args)
```

(範例程式碼：ch8/code05.py)

如果部署順利完成，就會看見該專案前方出現一個綠色打勾圖示，這時切換
到「觸發條件」頁籤，就可以看到所需要的 Webhook 網址。

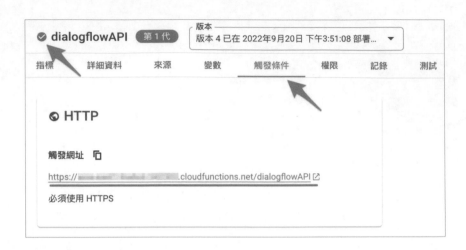

打開瀏覽器，在網址列輸入網址，後方加上輸入的文字參數，執行後就可以看見透過 Dialogflow 的回應訊息。

8-4　聊天機器人 (1) - LINE BOT 串接 Dialogflow (基本串接)

在前面三個小結裡，已經可以使用 DialogFlow 建立聊天機器人並串接後端伺服器，接著這篇教學會再繼續介紹，如何透過 Dialogflow 內建的功能，在不需要撰寫程式的狀態下，將聊天機器人的功能與 LINE BOT 串接，做出能夠理解自然語言的聊天機器人。

Dialogflow 建立 Intents 對話意圖、回覆內容

進入 Dialogflow 專案後，除了預設的 Defult Fallback Intent (未知的意圖) 和 Default Welcome Intent (歡迎意圖、打招呼意圖)，點擊上方 CREATE INTENT 建立新的 Intent，將其命名為 goodbye，並在 Training phrases 的區塊新增與「再見」相關的詞語。

如果某些詞彙出現 ENTITY 的對照，點擊後方的 X 刪除即可，因為某些詞彙可能會與內建的相同，Dialogflow 會自動進行串接 (貼心的行為？)，例如輸入「下次聊」，下次的「次」可能就會觸發。

相關語句建立完成後，接著往下建立 Responses 回覆內容，輸入一些相關的回覆語句，屆時 Dialogflow 會自動從中挑選進行回覆。

完成並儲存後，點擊右上的「Try it now」，就可以開始進行聊天，測試回覆結果 (因為 Dialogflow 會進行自然語言學習，所以就算輸入的字詞不是百分之百吻合，只要「意圖相同」，就會被歸類到相同的 Intent 裡)。

運用同樣的方式，可以建立好幾個「Intent 對話意圖」，完成後就可以準備開始串接 LINE BOT。

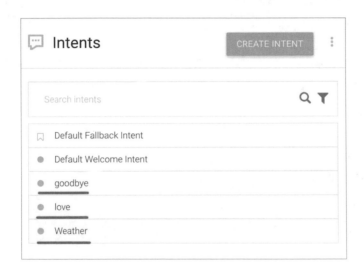

從 **Dialogflow** 建立與 **LINE** 串接的 **Webhook**

點擊左側「Integrations」頁籤,從中選擇 LINE。

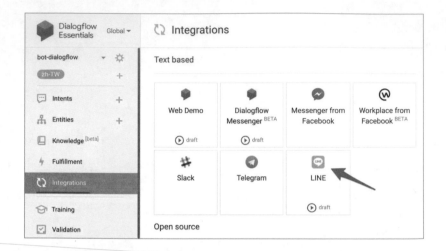

在彈出視窗裡,輸入 LINE 的相關資訊 (參考第二章「建立 LINE BOT」),
Dialogflow 就會產生一個 Webhook 網址,透過這個網址就能和 LINE 進行串接。

LINE BOT 串接 Dialogflow

前往 LINE Developer 控制台，進入欲串接 Dialogflow 的 LINE BOT，點擊 Message API 頁籤，勾選 Use webhook，將 Dialogflow 所產生的 Webhook 網址貼入 Webhook URL 的位置。

貼上網址後，按下 Verify 進行驗證，出現 Success 表示串接成功。

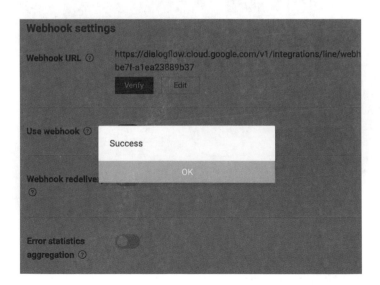

打開自己的 LINE，就可以開始與 LINE BOT 聊天，而 LINE BOT 進行自然語言辨識和回覆的內容，都是使用 Dialogflow 進行處理。

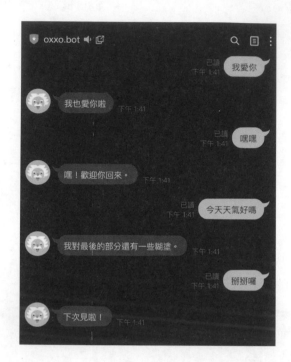

8-5 聊天機器人 (2) - LINE BOT 串接 Dialogflow (外部 Webhook)

延伸前一節的內容,接下來將額外讓 Dialogflow 串接 Python 建立的伺服器 Webhook,做到使用者與 LINE BOT 聊天時,不僅可以使用 Dialogflow 自然 語意分析的功能,也可以透過自己 Python 伺服器進行對應的邏輯處理。

串接流程圖

整個串接過程需要使用到「**兩個 Webhook 網址**」,從下方的串接流程圖, 可以了解如何使用 LINE 串接 Dialogflow 和自己的 Python 伺服器。

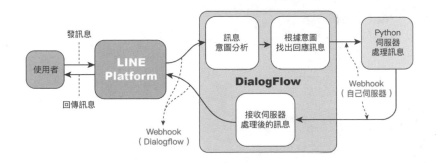

使用 Python 建立 Webhook 網址

參考「8-2、Dialogflow 串接 Webhook」章節,如果是透過「本機環境 + ngrok」或「Colab + ngrok」,可以執行下方程式碼產生測試的 Webhook (如果是 Colab,請參考「8-2、Dialogflow 串接 Webhook」裡的「建立 Webhook (Cloud Functions)」)。

```
# Colab 使用,本機環境請刪除
from flask_ngrok import run_with_ngrok
```

```python
from flask import Flask, request

app = Flask(__name__)

@app.route("/")
def home():
    return "<h1>hello world</h1>"

@app.route('/webhook', methods=['POST'])
def webhook():
    req = request.get_json()      # 轉換成 dict 格式
    print(req)
    reText = req['queryResult']['fulfillmentText']    # 取得回覆文字
    print(reText)
    return {
        "fulfillmentText": f'{reText} ( webhook )',
        "source": "webhookdata"
    }

# Colab 使用，本機環境請刪除
run_with_ngrok(app)

app.run()
```
(範例程式碼：ch8/code06.py)

程式執行後，開啟瀏覽器，輸入剛剛 ngrok 產生的網址 (記得先安裝並設定 ngrok，網址後方「不要」加上 /webhook)，如果出現 hello world 的文字，表示順利建立成功。

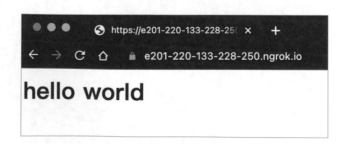

回到 Dialogflow 的專案裡，進入「Intents」頁籤，點擊需要串接 Webhook 的 Intent，**進入後在最下方勾選「Enable webhook call for this intent」，表示該 Intent 會透過 Webhook 處理後再進行回覆** (如果 Webhook 失敗則會直接套用內容回覆)。

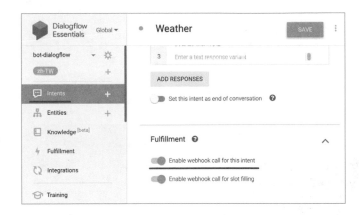

進入「Fulfillment」頁籤，勾選啟用 Webhook，將剛剛產生的 Webhook 貼上並儲存 (如果是本機環境或 Colab 網址，後方「要」加上 /webhook)

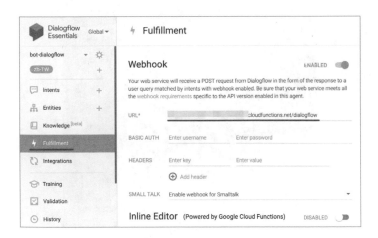

完成後在右側的聊天測試裡，輸入一些文字就可以看見機器人的自動回覆，如果回覆文字的後方有加上 (webhook) 文字，表示已經順利串接 Webhook (額外加上的文字是在 Webhook 伺服器端加入的，可以自行修改程式)。

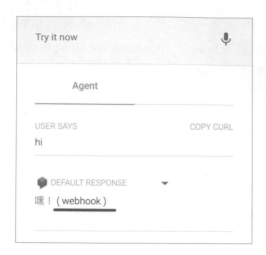

搭配 LINE BOT 測試

開啟 LINE 與 LINE BOT 的聊天視窗，輸入聊天內容，就可以看見已經透過自訂的 Webhook 網址進行回覆（回覆訊息後方會有自行添加的 webhook 文字）

Enable Webhook call for this intent

如果發現回傳的訊息，是沒有經過自己架設的 Python Webhook 伺服器 (例如最後一句「我愛你」的回覆後方並沒有 webhook)，表示該 Intent 沒有勾選 Enable Webhook call for this intent。

只要進到該 Intent 裡，將其勾選啟用，即可正常運作。

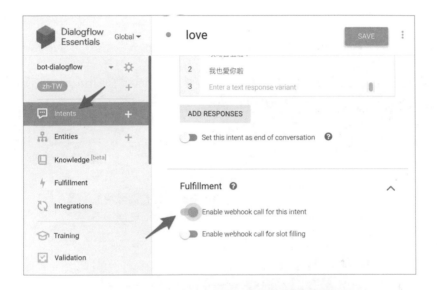

8-6 聊天機器人 (3) - LINE BOT 串接 Dialogflow (聊天問氣象)

這個章節會利用自己的 Webhook 伺服器解析 Dialogflow 與 LINE 傳遞的訊息，並將訊息透過 Requests 的方式直接傳送到 LINE，實作一個可以聊天，並透過聊天方式查詢雷達回波圖的簡單「氣象聊天機器人」。

建立 Dialogflow Intent

進入自己的 Dialogflow 專案，建立一個名為 radar 的 Intent，在 Training Phrases 的區塊輸入一些與「雷達回波圖」有關的詞彙。

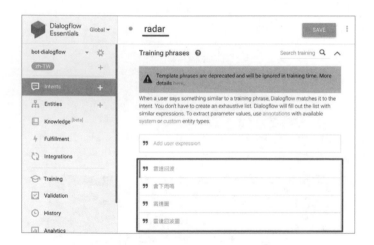

往下前往 **Responses 區塊**，輸入一個回應語句，因為待會會在 Webhook 伺服器端直接透過 Request 發送訊息，並不會使用到 Dialogflow 的回應，但因為串接需要有訊息內容，所以只需要一句即可，**輸入後往下勾選 nable Webhook call for this intent**。

回傳雷達回波圖 (本機環境)

前往自己本機環境的 Webhook 伺服器程式 (虛擬環境或 Anaconda Jupyter)，搭配下方的程式範例文章範例，輸入自己 LINE 的 Access Token，判斷當收到的 intent 為 radar 時，使用 requests 的方法直接回傳訊息 (詳細說明寫在程式碼內)。

> 由於 LINE Reply Token 只能使用一次，使用 request 回傳訊息後，就無法再透過 Dialogflow 傳送訊息，因此 return 回 Dialogflow 的訊息不會包含 fulfillmentText。

```python
from flask import Flask, request
import requests, json, time

app = Flask(__name__)

@app.route("/")
def home():
    return "<h1>hello world</h1>"
```

```
@app.route('/webhook', methods=['POST'])
def webhook():
    req = request.get_json()
    print(req)
    # 取得 Dialogflow 的回應文字
    reText = req['queryResult']['fulfillmentText']
    # 取得 intent 分類
    intent = req['queryResult']['intent']['displayName']
    # 取得 LINE replyToken
    replytoken = req['originalDetectIntentRequest']['payload']['data']
['replyToken']
    token = ' 你的 LINE BOT Access Token'
    # 雷達回波圖網址，後方加上時間戳記，避免緩存
    img = f'https://cwbopendata.s3.ap-northeast-1.amazonaws.com/MSC/
O-A0058-003.png?{time.time_ns()}'
    # 如果收到的 intent 是 radar
    if intent=='radar':
        headers = {'Authorization':'Bearer ' + token,'Content-
Type':'application/json'}
        body = {
            'replyToken':replytoken,
            'messages':[{
                    'type': 'image',
                    'originalContentUrl': img,
                    'previewImageUrl': img
                }]
            }
        # 使用 requests 方法回傳訊息到 L INE
        result = requests.request('POST', 'https://api.line.me/v2/bot/
message/reply',headers=headers,data=json.dumps(body).encode('utf-8'))
        print(result.text)
        # 完成後回傳訊息到 Dialogflow
        return {
            "source": "webhookdata"
        }
    # 如果收到的 intent 不是 radar
    else:
        # 使用 Dialogflow 產生的回應訊息
```

```
        return {
            "fulfillmentText": f'{reText} ( webhook )'
        }

app.run()
```
(範例程式碼：ch8/code07.py)

完成後啟動伺服器，除了可以在 LINE 裡與機器人聊天，也可以詢問雷達回
波圖。

自己的伺服器
使用 requests 方式回傳

Dialogflow + 自己的伺服器
由 Dialogflow 的 Webhook 回傳

回傳雷達回波圖 (Google Colab)

如果要使用 Colab 產生 Webhook，請先安裝 ngrok，接著使用下方的程式碼
(和本機環境的差別只在於有多了 flask_ngrok)

```
from flask import Flask, request
import requests, json, time
from flask_ngrok import run_with_ngrok
```

```python
app = Flask(__name__)

@app.route("/")
def home():
    return "<h1>hello world</h1>"

@app.route('/webhook', methods=['POST'])
def webhook():
    req = request.get_json()
    print(req)
    # 取得 Dialogflow 的回應文字
    reText = req['queryResult']['fulfillmentText']
    # 取得 intent 分類
    intent = req['queryResult']['intent']['displayName']
    # 取得 LINE replyToken
    replytoken = req['originalDetectIntentRequest']['payload']['data']['replyToken']
    token = ' 你的 Access Token'
    # 雷達回波圖網址，後方加上時間戳記，避免緩存
    img = f'https://cwbopendata.s3.ap-northeast-1.amazonaws.com/MSC/O-A0058-003.png?{time.time_ns()}'
    # 如果收到的 intent 是 radar
    if intent=='radar':
        headers = {'Authorization':'Bearer ' + token,'Content-Type':'application/json'}
        body = {
            'replyToken':replytoken,
            'messages':[{
                'type': 'image',
                'originalContentUrl': img,
                'previewImageUrl': img
            }]
        }
        # 使用 requests 方法回傳訊息到 LINE
        result = requests.request('POST', 'https://api.line.me/v2/bot/message/reply',headers=headers,data=json.dumps(body).encode('utf-8'))
        print(result.text)
        # 完成後回傳訊息到 Dialogflow
        return {
```

```
                "source": "webhookdata"
        }
    # 如果收到的 intent 不是 radar
    else:
        # 使用 Dialogflow 產生的回應訊息
        return {
            "fulfillmentText": f'{reText} ( webhook )'
        }


run_with_ngrok(app)       # 啟用 ngrok
app.run()
```
(範例程式碼：ch8/code08.py)

執行後，複製 Colab 裡 ngrok 所產生的網址。

```
 * Serving Flask app "__main__" (lazy loading)
 * Environment: production
   WARNING: This is a development server. Do not use it in a production deployment.
   Use a production WSGI server instead.
 * Debug mode: off
INFO:werkzeug: * Running on http://127.0.0.1:5000/ (Press CTRL+C to quit)
 * Running on http://c963-35-230-74-66.ngrok.io
 * Traffic stats available on http://127.0.0.1:4040
```

將網址貼回 Dialogflow 的 Fulfillment Webhook 網址裡 (注意，要改成 https)

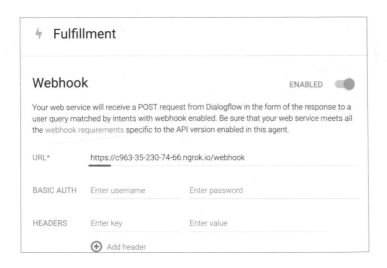

完成後就可以和 LINE 機器人聊天，詢問雷達回波圖。

自己的伺服器
使用 requests 方式回傳

Dialogflow + 自己的伺服器
由 Dialogflow 的 Webhook 回傳

回傳雷達回波圖 (Cloud Functions)

透過 Google Cloud Functions 可以建立一個可以 24 小時運作的 Webhook 網址，在程式碼編輯的 main.py 輸入下方的程式碼。

```python
import requests, json, time
def webhook(request):
    try:
        req = request.get_json()
        reText = req['queryResult']['fulfillmentText']
        intent = req['queryResult']['intent']['displayName']
        replytoken = req['originalDetectIntentRequest']['payload']
['data']['replyToken']
        token = ' 你的 LINE BOT Access Token'
        img = f'https://cwbopendata.s3.ap-northeast-1.amazonaws.com/
MSC/O-A0058-003.png?{time.time_ns()}'
        if intent=='radar':
            headers = {'Authorization':'Bearer ' + token,'Content-
Type':'application/json'}
```

```python
        body = {
            'replyToken':replytoken,
            'messages':[{
                'type': 'image',
                'originalContentUrl': img,
                'previewImageUrl': img
            }]
        }
    result = requests.request('POST', 'https://api.line.
me/v2/bot/message/reply',headers=headers,data=json.dumps(body).
encode('utf-8'))
        print(result.text)
        return {
            "source": "webhookdata"
        }
    else:
        return {
            "fulfillmentText": f'{reText} ( webhook )'
        }
    except:
        print(request.args)
```

(範例程式碼：ch8/code09.py)

因為 requests 為第三方函式庫，所以需要在 requirements.txt 裡添加
requests。

完成後如果成功部署，會出現綠色打勾圖示，複製觸發條件的 Webhook 網
址。

回到 Dialogflow 的 Fulfillment 頁籤，將網址貼到 Webhook 的 URL 欄位。

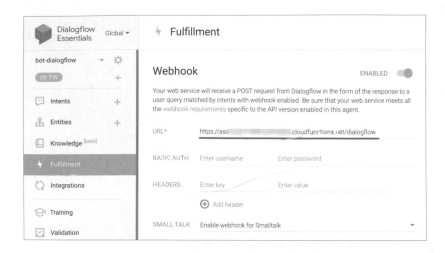

完成並儲存後，與 LINE BOT 聊天時，就會透過 Google Cloud Functions 進行邏輯判斷，並回傳雷達回波圖。

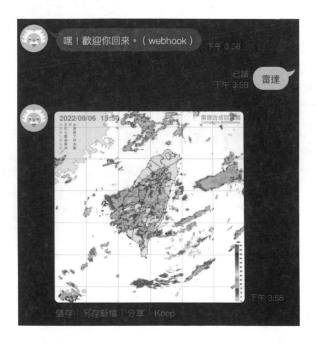

8-7 聊天機器人 (4) - LINE BOT 串接 Dialogflow (接收表情貼圖)

在之前「LINE BOT 串接 Dialogflow」的三個章節裡，都是使用 Dialogflow 產生的 Webhook 串接 LINE 機器人，但因為 Dialogflow 只能處理「文字」，如果遇到「表情貼圖」、「地圖資訊」等非文字訊息，就會發生無法處理的狀況。接下來將會介紹如何透過自建的 Webhook 伺服器解析 LINE 訊息後，再將文字交由 Dialogflow 處理，讓串接 Dialogflow 的 LINE BOT 可以正確處理聊天訊息。

伺服器串接 Dialogflow 流程圖

架設自己的 Python 伺服器後，會透過 Webhook 網址與 LINE BOT 連結，透過 Dialogflow 的 API 和 Dialogflow 連接，彼此之間連動的關係，可以參考下方的串接流程圖：

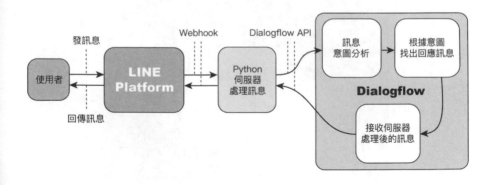

接收表情貼圖 (本機環境)

參考前幾個章節的文章範例，使用下方的程式碼，將 LINE BOT 的 Webhook 結合 Dialogflow API，在收到 LINE 訊息時，**解析訊息的類型 type**，如果

type 為 text 表示純文字，就將訊息內容提供給 Dialogflow 進行自然語言處理，如果不是純文字，則直接回傳「你傳的不是文字呦」的訊息（要填入自己 LINE BOT 的 Access Token 以及 Channel Secret，詳細說明寫在程式碼的註解中）。

> 注意，因 Google Dialogflow 函式庫無法運行在 Python 3.7 的環境，所以如果遇到無法安裝的情形，請先將 Python 升級為 3.9 以上版本，同理，因為 Colab 預設 Python 3.7，也就無法正確安裝和執行 Google Dialogflow 函式庫。

```
import os
import google.cloud.dialogflow_v2 as dialogflow
from flask import Flask, request

# 載入 json 標準函式庫，處理回傳的資料格式
import json

# 載入 LINE Message API 相關函式庫
from linebot import LineBotApi, WebhookHandler
from linebot.exceptions import InvalidSignatureError
from linebot.models import MessageEvent, TextMessage, TextSendMessage

os.environ["GOOGLE_APPLICATION_CREDENTIALS"] = 'dialogflow_key.json' #
金鑰 json
project_id = 'XXXX'          # dialogflow project id
language = 'zh-TW'           # 語系
session_id = 'oxxostudio'    # 自訂 session id

# dialogflow 處理自然語言
def dialogflowFn(text):
    session_client = dialogflow.SessionsClient()
    session = session_client.session_path(project_id, session_id)
    text_input = dialogflow.types.TextInput(text=text, language_
code=language)
    query_input = dialogflow.types.QueryInput(text=text_input)
    print(query_input)
```

```python
    try:
        response = session_client.detect_intent(session=session,
query_input=query_input)
        print("input:", response.query_result.query_text)
        print("intent:", response.query_result.intent.display_name)
        print("reply:", response.query_result.fulfillment_text)
        return response.query_result.fulfillment_text
    except:
        return 'error'


app = Flask(__name__)


@app.route("/", methods=['POST'])
def linebot():
    # 取得收到的訊息內容
    body = request.get_data(as_text=True)
    try:
        # json 格式化訊息內容
        json_data = json.loads(body)
        access_token = 'Access Token'
        secret = 'Channel Secret'
        # 確認 token 是否正確
        line_bot_api = LineBotApi(access_token)
        # 確認 secret 是否正確
        handler = WebhookHandler(secret)
        # 加入回傳的 headers
        signature = request.headers['X-Line-Signature']
        # 綁定訊息回傳的相關資訊
        handler.handle(body, signature)
         # 取得回傳訊息的 Token
        tk = json_data['events'][0]['replyToken']
        # 取得 LINe 收到的訊息類型
        type = json_data['events'][0]['message']['type']
        # 取得 LINE 收到的文字訊息
        if type=='text':
            msg = json_data['events'][0]['message']['text']
            # 印出內容
            print(msg)
            # dialogflow 處理後回傳文字
```

```
        reply = dialogflowFn(msg)
    else:
        reply = '你傳的不是文字呦～'
    print(reply)
    # 回傳訊息
    line_bot_api.reply_message(tk,TextSendMessage(reply))
except:
    print(body)
# 驗證 Webhook 使用，不能省略
return 'OK'

if __name__ == "__main__":
    app.run()
```
(範例程式碼：ch8/code10.py)

完成後使用 ngrok 產生的 Webhook，將網址填入 LINE Developer 裡，完成
後與 LINE 機器人聊天，機器人就可以識別表情貼圖或文字，如果是文字，
就會透過 Dialogflow 處理並回傳結果。

接收表情貼圖 (Cloud Functions)

進入 Cloud Functions 專案後，加入 dialogflow 的 API 金鑰 json，並在 requirements.txt 裡新增下列三個第三方函式庫 (詳細操作步驟可以參考 8-3 裡的「串接 Dialogflow (Cloud Functions)」)。

```
google-cloud-dialogflow
line-bot-sdk
requests
```

輸入下方的程式碼，完成後點擊下方的「部署」，就會將程式部署到 Cloud Functions 裡 (要填入自己 LINE BOT 的 Access Token 以及 Channel Secret)。

```python
import os
import google.cloud.dialogflow_v2 as dialogflow

import json
from linebot import LineBotApi, WebhookHandler
from linebot.exceptions import InvalidSignatureError
from linebot.models import MessageEvent, TextMessage, TextSendMessage

# 剛剛下載的金鑰 json
os.environ["GOOGLE_APPLICATION_CREDENTIALS"] = 'dialogflow_key.json'
project_id = 'XXXXX'          # dialogflow 的 project id
language = 'zh-TW'            # 語系
```

```python
session_id = 'oxxostudio'    # 自訂義的 session id

def dialogflowFn(text):
    # 使用 Token 和 dialogflow 建立連線
    session_client = dialogflow.SessionsClient()
    # 連接對應專案
    session = session_client.session_path(project_id, session_id)
    # 設定語系
    text_input = dialogflow.types.TextInput(text=text, language_
code=language)
    # 根據語系取得輸入內容
    query_input = dialogflow.types.QueryInput(text=text_input)
    try:
        # 連線 Dialogflow 取得回應資料
        response = session_client.detect_intent(session=session,
query_input=query_input)
        print("input:", response.query_result.query_text)
        print("intent:", response.query_result.intent.display_name)
        print("reply:", response.query_result.fulfillment_text)
        # 回傳回應的文字
        return response.query_result.fulfillment_text
    except:
        return 'error'

def webhook(request):
    # 取得收到的訊息內容
    body = request.get_data(as_text=True)
    try:
        # json 格式化訊息內容
        json_data = json.loads(body)
        access_token = 'Access Token'
        secret = 'Channel Secret'
        # 確認 token 是否正確
        line_bot_api = LineBotApi(access_token)
        # 確認 secret 是否正確
        handler = WebhookHandler(secret)
        # 加入回傳的 headers
        signature = request.headers['X-Line-Signature']
        # 綁定訊息回傳的相關資訊
```

```
    handler.handle(body, signature)
    # 取得回傳訊息的 Token
    tk = json_data['events'][0]['replyToken']
    # 取得 LINe 收到的訊息類型
    type = json_data['events'][0]['message']['type']
    if type=='text':
        # 取得 LINE 收到的文字訊息
        msg = json_data['events'][0]['message']['text']
        # 印出內容
        print(msg)
        reply = dialogflowFn(msg)
    else:
        reply = ' 你傳的不是文字呦～'
    print(reply)
    # 回傳訊息
    line_bot_api.reply_message(tk,TextSendMessage(reply))
except:
    print(body)
# 驗證 Webhook 使用，不能省略
return 'OK'
```

（範例程式碼：ch8/code11.py）

完成後如果成功部署，會出現綠色打勾圖示，複製觸發條件的 Webhook 網址。

將 Webhook 網址填入 LINE Developer 裡，完成後與 LINE 機器人聊天，機器人就可以識別表情貼圖或文字，如果是文字，就會透過 Dialogflow 處理並回傳結果。

 小結

過去在實作 LINE 的聊天機器人時，最麻煩的就是處理「語句的判斷」，但如果將聊天內容交由 Dialogflow 進行分析處理判斷，並進一步將 LINE 與 Dialogflow、自訂的 Python 伺服器串接，就能更為準確的處理自然語言（使用者自己的語意），做到更逼真的 LINE 聊天機器人了。

Note

9

使用 LINE Notify
推播通知

前言

除了可以開發自己的 LINE 機器人，也可以使用 LINE 內建的 LINE Notify 服務，單純的進行推播通知，有別於 LINE 機器人可以進行聊天互動，雖然 LINE Notify 只能在觸發某些事件後發送通知，但相對開發一個 LINE 機器人來說簡單許多，接下來會介紹如何使用 LINE Notify。

> **本章節的範例程式碼：**
>
> https://github.com/oxxostudio/book-code/tree/master/linebot/ch9

9-1 認識 LINE Notify

LINE Notify 是 LINE 所提供的一項非常方便的服務,**用戶可以透過 LINE,接收各種網站、服務或應用程式 (GitHub、IFTTT 及 Python... 等) 的提醒通知**,與網站服務連動完成後,LINE 所 提供的官方帳號「LINE Notify」將會傳送通知,不僅可與多個服務連動,也可透過 LINE 群組接收通知。

> LINE Notify 網址:https://notify-bot.line.me/zh_TW/

9-2　申請 LINE Notify 權杖

打開 LINE Notify 的網站後，使用自己的 LINE 帳號登入，登入後從上方個人
帳號，選擇「個人頁面」。

進入個人頁面後，點選下方「發行權杖」，**權杖 (token) 的作用在於讓「連
動的服務」可以透過 LINE Notify 發送訊息通知。**

點選「發行權杖」後，必須要定義權杖的名稱，以及選擇這個 LINE Notify
所在的聊天群組，通常直接選擇「**透過 1 對 1 聊天接收 LINE Notify 通知**」
(如果將 LINE Notify 加入群組，群組中所有的成員都會收到推播通知)。

發行權杖後，會出現一串權杖代碼，點擊下方綠色的「複製」就可複製權杖代碼。

> 注意，權杖代碼只會出現一次，複製後自行找地方留存。

點擊關閉,在個人頁面裡就會看見已經發行的權杖,點選後方「刪除」就能解除權杖(如果不小心權杖流出導致一直收到奇怪的通知,就可以將權杖解除,重新再發行一次)。

權杖發行後,在個人的 LINE 裡,就會收到「已發行個人權杖」的通知訊息(解除權杖也會收到通知)。

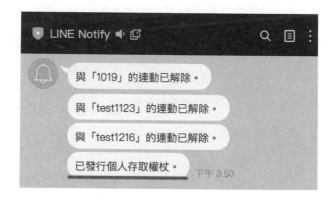

9-3　發送 LINE Notify 訊息

有了 LINE Notify 的權杖後，就能使用 Requests 的 POST 方法發送訊息，使用 Colab 或本機環境（全新的 Python 虛擬環境需要使用 pip install requests 安裝，Anaconda Jupyter 和 Colab 已經內建，不用安裝），發送時需要在 headers 設權杖 Authorization，並將訊息內容放在 data 的 message 裡。

> **Requests 教學：**
>
> https://steam.oxxostudio.tw/category/python/spider/requests.html

```
import requests

url = 'https://notify-api.line.me/api/notify'
token = '剛剛複製的權杖'
headers = {
    'Authorization': 'Bearer ' + token     # 設定權杖
}
data = {
    'message':'測試一下！'        # 設定要發送的訊息
}
# 使用 POST 方法發送訊息
data = requests.post(url, headers=headers, data=data)
```
（範例程式碼：ch9/code01.py）

執行 Python 程式，就會在 LINE 裡就收到 LINE Notify 的推播通知。

9-4 透過 LINE Notify 發送表情貼圖

前往 LINE 官方表情貼圖清單網頁，查看 stickerPackageId (貼圖包類別號碼)
和 stickerId (貼圖號碼)。

> **表情貼圖清單：**
>
> https://developers.line.biz/en/docs/messaging-api/sticker-list/

修改剛剛 LINE Notify 發送訊息的程式，在 data 裡加入 stickerPackageId 和
stickerId，就能夠發送表情貼圖。

```python
import requests

url = 'https://notify-api.line.me/api/notify'
token = '剛剛複製的權杖'
headers = {
    'Authorization': 'Bearer ' + token    # 設定權杖
}
```

```
data = {
    'message':' 測試一下！'        # 設定要發送的訊息
}
# 使用 POST 方法發送訊息
data = requests.post(url, headers=headers, data=data)
```
(範例程式碼：ch9/code02.py)

9-5 透過 LINE Notify 傳送圖片

修改程式碼，在發送的 data 裡，加入 imageThumbnail（縮圖網址）和
imageFullsize（圖片網址），就能夠傳送圖片。

```python
import requests

url = 'https://notify-api.line.me/api/notify'
token = ' 剛剛複製的權杖 '
headers = {
    'Authorization': 'Bearer ' + token
}
data = {
    'message':' 測試一下！',
    'imageThumbnail':'https://steam.oxxostudio.tw/downlaod/python/
line-notify-demo.png',
    'imageFullsize':'https://steam.oxxostudio.tw/downlaod/python/line-
notify-demo.png'
}
data = requests.post(url, headers=headers, data=data)
```
(範例程式碼：ch9/code03.py)

9-6 透過 LINE Notify 傳送雷達回波圖

了解 LINE Notify 的用法後，接下來會使用 Python 的 Requests 函式庫，爬取政府開放資料裡的氣象局雷達回波圖，並結合 LINE Notify 的 API，透過 LINE 傳送雷達回波圖。

取得雷達回波圖網址

雷達回波圖除了可以直接從中央氣象局的網站取得，在政府資料開放平臺也有提供即時的圖片資料，在平台的資料集裡點擊連結後會下載一份 JSON，下載後使用記事本或是 chrome 瀏覽器開啟檔案，開啟後可以看到檔案內「cwbopendata > dataset > resource > uri」的位置是一張圖片網址，這張圖片就是所需要的雷達回波圖。下方列出常用的兩種雷達回波圖。

> 關於雷達回波圖：https://data.gov.tw/dataset/75125

資料資源下載網址		
↓ XML	檢視資料	雷達整合回波透明圖層(鄰近區域)
↓ JSON	檢視資料	雷達整合回波透明圖層(鄰近區域)
↓ JSON	檢視資料	雷達整合回波透明圖層(較大範圍)
↓ XML	檢視資料	雷達整合回波透明圖層(較大範圍)
↓ JSON	檢視資料	臺灣(鄰近區域)_有地形
↓ JSON	檢視資料	臺灣(較大範圍)_有地形
↓ JSON	檢視資料	臺灣(鄰近區域)_無地形
↓ JSON	檢視資料	臺灣(較大範圍)_無地形
↓ XML	檢視資料	臺灣(鄰近區域)_有地形

```
{
"cwbopendata": {
  "@xmlns": "urn:cwb:gov:tw:cwbcommon:0.1",
  "identifier": "371855a8-d9a5-43fa-8017-0805e6f15776",
  "sender": "weather@cwb.gov.tw",
  "sent": "2021-12-09T00:26:38+08:00",
  "status": "Actual",
  "msgType": "Issue",
  "scope": "Public",
  "dataid": "A0058-003",
  "source": "Meteorological Satellite Center",
  "dataset": {
   "datasetInfo": {
    "datasetDescription": "資料說明",
    "parameterSet": {
     "parameter": [
       {
        "parameterName": "整合雷達名",
        "radarName": "五分山、花蓮、七股、墾丁、樹林、南屯、林園雷達"
       },
       {
        "parameterName": "經度範圍",
        "parameterValue": "118.0-124.0"
       },
       {
        "parameterName": "緯度範圍",
        "parameterValue": "20.5-26.5"
       },
       {
        "parameterName": "解析度",
        "parameterValue": "3600x3600"
       }
      ]
     }
    },
    "resource": {
     "resourceDesc": "雷達整合回波圖-臺灣(鄰近地區)_無地形",
     "mimeType": "image/png",
     "uri": "https://cwbopendata.s3.ap-northeast-1.amazonaws.com/MSC/O-A0058-003.png"
    },
    "time": {
     "obsTime": "2021-12-09T00:20:00+08:00"
```

下方列出常用的兩種雷達回波圖網址，由於圖片網址不會變動，可以直接在程式碼內使用。

- 臺灣 (鄰近區域)_ 無地形：

  https://cwbopendata.s3.ap-northeast-1.amazonaws.com/MSC/O-A0058-001.png

- 臺灣 (較大範圍)_ 無地形：

  https://cwbopendata.s3.ap-northeast-1.amazonaws.com/MSC/O-A0058-003.png

LINE Notify 傳送雷達回波圖

使用 requests 函式庫，爬取剛剛開發資料雷達回波圖的 JSON 網址，將抓到的資料轉換為 JSON 格式，從中取得圖片網址，取得網址後，再次使用 requests 函式庫的 POST 的方法，透過 LINE Notify 將雷達回波圖傳送到 LINE 裡。

Requests 教學：

https://steam.oxxostudio.tw/category/python/spider/requests.html

---

```python
import requests

url = 'https://notify-api.line.me/api/notify'
# 自己申請的 LINE Notify 權杖
token = '你的 LINE Notify 權杖'
# POST 使用的 headers
headers = {
    'Authorization': 'Bearer ' + token
}
# POST 使用的 data
data = {
    'message':' 從雷達回波看看會不會下雨～ ',
    'imageThumbnail':'https://cwbopendata.s3.ap-northeast-1.amazonaws.
com/MSC/O-A0058-003.png',
    'imageFullsize':'https://cwbopendata.s3.ap-northeast-1.amazonaws.
com/MSC/O-A0058-003.png'
}
data = requests.post(url, headers=headers, data=data)
```
（範例程式碼：ch9/code04.py）

避免緩存 (Cache) 變成舊圖片

實作過程中，可能會遇到如果傳送過圖片，過一陣子再次傳送時，顯示的是「舊」的雷達回波圖，而非「即時」雷達回波圖，這是**因為 LINE 認為「同一個網址」就應該是「同一張圖片」(API 裡的圖片網址不會變，只有圖片內容變了)，所以在傳送時可以替圖片加上「時間參數」，就能避開這個問題。**

修改剛剛的程式，在發送的圖片後方，加上現在的時間資訊（使用標準函式 time.time_ns()），在傳送時就能讓 LINE 認為是不同的圖片，並能顯示正確的圖片了。

```
import requests

url = 'https://notify-api.line.me/api/notify'
# 自己申請的 LINE Notify 權杖
token = ' 你的 LINE Notify 權杖 '
# POST 使用的 headers
headers = {
    'Authorization': 'Bearer ' + token
}
# POST 使用的 data
data = {
    'message':' 從雷達回波看看會不會下雨～ ',
    'imageThumbnail':'https://cwbopendata.s3.ap-northeast-1.amazonaws.
com/MSC/O-A0058-003.png',
    'imageFullsize':'https://cwbopendata.s3.ap-northeast-1.amazonaws.
com/MSC/O-A0058-003.png'
}
data = requests.post(url, headers=headers, data=data)
```

 小結

除了使用 LINE 機器人進行互動之外，也可以透過 LINE Notify 進行單純的「推播」，在不需要撰寫大量的程式碼的狀況下，也能透過 LINE 進行基本的通知，是相當實用且方便的功能。

Note

10

使用 Google Cloud Functions

由於在開發 LINE BOT 的過程中，如果要建構 24 小時的服務，往往會使用到 Google Cloud Functions，這個章節會介紹如何使用使用 Google Cloud Functions。

本章節的範例程式碼：

https://github.com/oxxostudio/book-code/tree/master/linebot/ch10

10-1 認識 Google Cloud Functions

Google Cloud Functions 是 Google Cloud 裡的服務，一個無伺服器的雲端執行環境，可以部署一些簡單或單一用途的程式，當監聽的事件被觸發時，就會觸發 Cloud Function 裡所部署的程式，由於不需要伺服器的特性，常常作為輕量化的 API 以及 Webhook 使用。

前往 Google Cloud Functions：https://cloud.google.com/functions

Google Cloud Functions 可 以 使 用 Node.js、Python、Go、Java、.NET 或 Ruby 程式語言進行編輯，執行環境也會因為選擇的運行時而產生差異，針對伺服器管理、軟體設置、框架更新或作業系統更新等基礎設施的建置，全都由 Google 負責管理，使用者完全不需要做任何事情，只需要專注在自己的程式碼邏輯。

10-2 Cloud Functions 計費方式

啟用 Cloud Functions 需要綁定個人信用卡,但 Cloud Functions 有提供免費的額度供開發者使用,如果用量不超過額度,則第一年內不會收費(Google 提供第一年幾百美金的額度,通常單純個人操作或小型的應用開發,基本上不可能一天百萬次的呼叫額度),第二年開始的計費依據則會以函式的執行時間長度、函式的叫用次數,以及您為函式佈建的資源數量來計算,最低一個月約 0.01 美金。

參考 Cloud Functions 定價:

https://xn--cloud-b54d7f.google.com/functions/pricing

10-3 建立 Cloud Cloud 專案

前往 Google Cloud Functions 頁面，點擊「免費試用」。

前往 Google Cloud Functions：

https://cloud.google.com/functions

Cloud Functions

可靈活擴充的函式即服務 (FaaS)，讓您以即付即用的方式執行程式碼，完全不必管理伺服器。

免費試用

第一步，下拉選擇個人基本資訊，勾選同意服務條款。

第二步，輸入電話號碼，取得驗證碼後輸入。

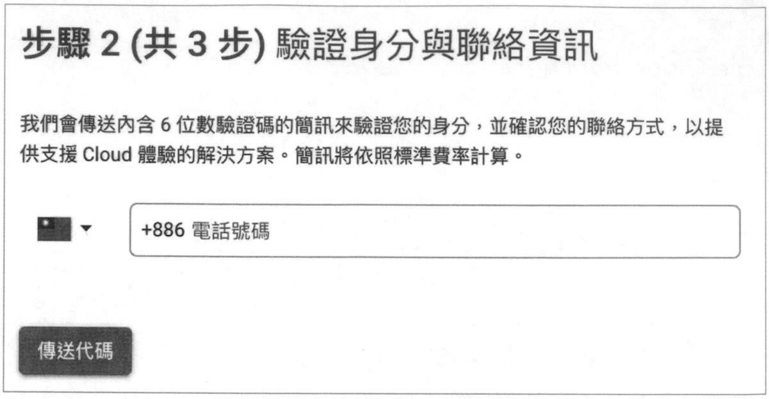

步驟 2 (共 3 步) 驗證身分與聯絡資訊

我們會傳送內含 6 位數驗證碼的簡訊來驗證您的身分，並確認您的聯絡方式，以提供支援 Cloud 體驗的解決方案。簡訊將依照標準費率計算。

🇹🇼 ▼　　　+886 電話號碼

傳送代碼

第三步，輸入個人住址、信用卡資訊，點擊開始免費試用，就能開始使用。

步驟 3 (共 3 步) 驗證付款資訊

您的付款資訊可協助我們減少欺詐活動和濫用行為。除非開啟自動計費功能，否則系統不會向您收取任何費用。

付款方式 ⓘ

\#　　信用卡詳細資料

☑ 信用卡或簽帳金融卡地址同上

完成後，進入 Google Cloud Platform（GCP）控制台，建立一個專案，如果沒有建立過專案，可能會要求建立一個新專案。

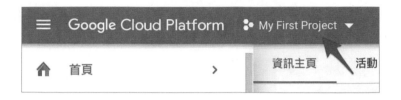

10-4　啟用 Cloud Build API

因為 Cloud Functions 需要搭配 Cloud Build API，必須先啟用 Cloud Build
API，建立專案之後，點擊左上方圖示開啟左側選單，選擇「API 和服務 >
程式庫」。

從程式庫裡搜尋「build api」。

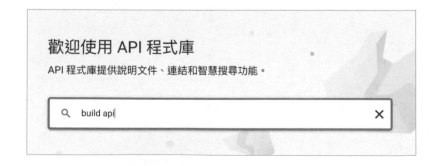

搜尋到 Build API 後，點擊「啟用」。

接著點擊「啟用計費功能」。

選擇付款的帳戶後，就可以啟用 Cloud Build API（不需要太過擔心付費的問題，因為如果是個人用戶的用量，第一年基本上完全免費）。

完成後會出現「API」已啟用的標示，表示啟用完成。

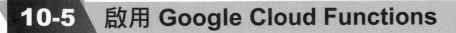

10-5 啟用 Google Cloud Functions

專案建立完成後，如果沒有自動跳轉到 Cloud Functions，點選左上角圖示展開選單，選擇 Cloud Functions。

點選「建立函式」，就能開始建立第一支 Cloud Functions 的程式。

10 使用 Google Cloud Functions

點選建立函式後,設定基本資訊,環境選擇「第一代」(因為第二代在這個時間點貌似在 beta 階段),函式名稱自行定義,區域選擇「asia-east1」台灣主機(理論上速度比較快)。

基本資訊

環境
第 1 代

函式名稱 *
test

區域
asia-east1

觸發條件選擇「HTTP」,勾選「允許未經驗證的叫用」,如此一來就能單純透過 request POST 或 GET 的方法進行叫用,也符合大多數聊天機器人的叫用機制。

🌐 HTTP

觸發條件類型
HTTP

網址 📋

https://asia-████-████████.cloudfunctions.net/test

驗證
◉ 允許未經驗證的叫用
　如要建立公用 API 或網站,請勾選這個選項。

○ 需要驗證
　利用 Cloud IAM 管理已獲授權的使用者。

☑ 必須使用 HTTPS ❓

[儲存]　取消

執行階段、建構作業、連線和安全型設定則不需要更動，保留預設值即可。

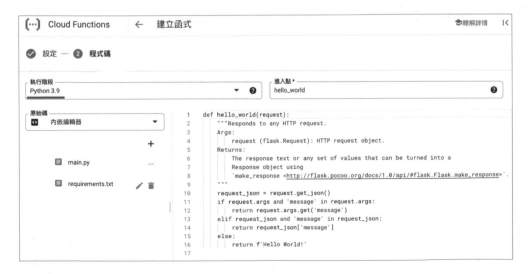

點選下一步，進入程式碼編輯畫面，從左上方下拉選單選擇 Python（3.7～3.9 皆可），程式編輯區就會自動出現預設的 Python 樣板，就可以準備開始編輯程式。

10-6 部署第一支程式

畫面左側目錄裡 main.py 表示這支程式的主程式碼，requirements.txt 表示要額外安裝的外部函式庫。

右上方的「進入點」表示使用 HTTP 的方式叫用這支程式時，要執行的 funciton，也就是下方程式碼的 hello_world。

```
                     ▼   ❓      進入點 *
                                hello_world                                              ❓

1    def hello_world(request):
2        """Responds to any HTTP request.
3        Args:
4            request (flask.Request): HTTP request object.
5        Returns:
6            The response text or any set of values that can be turned into a
7            Response object using
8            `make_response <http://flask.pocoo.org/docs/1.0/api/#flask.Flask.make_response>`.
9        """
10       request_json = request.get_json()
11       if request.args and 'message' in request.args:
12           return request.args.get('message')
13       elif request_json and 'message' in request_json:
14           return request_json['message']
15       else:
16           return f'Hello World!'
17
```

在完全不執行任何動作的情況下（使用預設 Python 樣板），點擊下方的「部署」，就能部署最基本的網頁應用程式。

部署完成後，前方會出現綠色打勾的圖示，表示部署一切正常（沒有發生任何錯誤）。

點擊進入程式，選擇「觸發條件」，點擊並開啟網址，就會看見網頁出現「Hello World!」，表示已經可以透過外部呼叫所部署的程式。

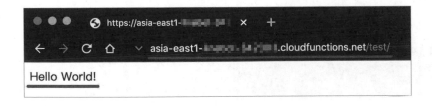

10-7 讀取參數

由於 Cloud Functions 的 Python 是使用 Flask 函式庫的架構，因此讀取參數的方式和 Flask 相同，將原本的程式碼改成下方的內容，修改並重新部署完成後，輸入網址叫用程式後，從「記錄」裡，就能看到叫用時對應的參數資訊。

```python
def hello_world(request):
    request_json = request.get_json()
    print(request.args )     # 讀取 GET 方法參數
    print(request.form )     # 讀取 POST 方法參數
    print(request.path )     # 讀取網址
    print(request.method)    # 讀取叫用方法
    if request.args and 'message' in request.args:
        return request.args.get('message')
    elif request_json and 'message' in request_json:
        return request_json['message']
    else:
        return f'Hello World!'
```

(範例程式碼：appendix/code01.py)

> ⚙	2022-03-02 15:41:19.823 台北	test	ksgt82qmoueq	⬚	Function execution took 6 ms, finished wit
> ⚙	2022-03-02 15:41:31.240 台北	test	ksgt1k0k9a69	⬚	Function execution started
> ✳	2022-03-02 15:41:31.243 台北	test	ksgt1k0k9a69	⬚	ImmutableMultiDict([('message', '789')])
> ✳	2022-03-02 15:41:31.243 台北	test	ksgt1k0k9a69	⬚	ImmutableMultiDict([])
> ✳	2022-03-02 15:41:31.243 台北	test	ksgt1k0k9a69	⬚	/
> ✳	2022-03-02 15:41:31.243 台北	test	ksgt1k0k9a69	⬚	GET

如果使用後端程式，亦可發出 POST 的請求。

```
import requests
data = {'name': 'oxxo', 'age': '18'}
web = requests.post('https://asia-east1-XXXXXX.cloudfunctions.net/
test', data=data)   # 發送 POST 請求
print(web.text)
```

>	✿	2022-03-02 15:46:16.703 台北	test	9e8w9k8s0o45	≋	Function execution started
>	✳	2022-03-02 15:46:16.705 台北	test	9e8w9k8s0o45	≋	ImmutableMultiDict([])
>	✳	2022-03-02 15:46:16.707 台北	test	9e8w9k8s0o45	≋	ImmutableMultiDict([('name', 'oxxo'), ('age', '18')])
>	✳	2022-03-02 15:46:16.707 台北	test	9e8w9k8s0o45	≋	/
>	✳	2022-03-02 15:46:16.707 台北	test	9e8w9k8s0o45	≋	POST
>	✿	2022-03-02 15:46:16.708 台北	test	9e8w9k8s0o45	≋	Function execution took 6 ms, finished with status code: 200

10-8　處理跨域問題

透過瀏覽器開啟網頁呼叫 API 時，常常會遭遇「跨域」的問題（因為瀏覽器的安全性限制，不同網域間無法直接叫用），使用 Cloud Functions 建立的 API 預設禁止跨域叫用，但只要加入下方的程式碼，就能夠允許跨域叫用。

```python
def hello_world(request):
    request_json = request.get_json()
    print(request.args )     # 讀取 GET 方法參數
    print(request.form )     # 讀取 POST 方法參數
    print(request.path )     # 讀取網址
    print(request.method)    # 讀取叫用方法

    headers = {
        'Access-Control-Allow-Origin': '*',
        'Access-Control-Allow-Headers': 'Content-Type',
        'Access-Control-Max-Age': '3600'
    }

    return ('Hello World!', 200, headers)  # 回傳同意跨域的 header
```
（範例程式碼：appendix/code02.py）

完成後，在網頁端執行對應的 JavaScript，就能得到正確的結果（下方程式碼為 JavaScript）。

```javascript
let uri = ' 你的網址 ';
fetch(uri, {method:'GET'})
.then(res => {
    return res.text()
}).then(result => {
    console.log(result);
});
```

 小結

Google Cloud Functions 是一個非常方便的 Google 雲端服務，如果不需要複雜的後端程式 (例如使用大量的 GPU 或 CPU 運算或資料庫)，可以單純藉由 Google Cloud Functions 創造輕量化的 Web API，做出各種有趣的網路應用或串接各式各樣的網路服務。

Note

Note

附錄

其他參考資訊

前 言

只要了解了 LINE BOT 的運作原理，要實作一個客製化的 LINE BOT 其實不困難，然而 Python 博大精深，在開發過程中，常常會需要使用一些 Python 程式技巧或函式庫，因此透過最後的附錄整理的網址，列出本書會使用到的一些相關語法，掌握這些程式語法後，就能更清楚掌握 LINE BOT 的開發技巧。

Python 基本資料型別

- 變數 variable：
  https://steam.oxxostudio.tw/category/python/basic/variable.html

- 數字 number：
  https://steam.oxxostudio.tw/category/python/basic/number.html

- 文字與字串 string：
  https://steam.oxxostudio.tw/category/python/basic/string.html

- 串列 list：
  https://steam.oxxostudio.tw/category/python/basic/list.html

- 字典 dictionary：
  https://steam.oxxostudio.tw/category/python/basic/dictionary.html

- 元組 (數組) tuple：
  https://steam.oxxostudio.tw/category/python/basic/tuple.html

- 集合 set：
  https://steam.oxxostudio.tw/category/python/basic/set.html

Python 重要的基本語法

- 縮排和註解：
  https://steam.oxxostudio.tw/category/python/basic/ident.html

- 邏輯判斷：
  https://steam.oxxostudio.tw/category/python/basic/if.html

- 重複迴圈：
  https://steam.oxxostudio.tw/category/python/basic/loop.html

- 函式 function：
  https://steam.oxxostudio.tw/category/python/basic/function.html

Python 常用的函式庫 (模組)

- Flask 函式庫：
  https://steam.oxxostudio.tw/category/python/example/flask.html

- Requests 函式庫：
  https://steam.oxxostudio.tw/category/python/spider/requests.html

- JSON 檔案操作：
  https://steam.oxxostudio.tw/category/python/library/json.html

Deepen Your Mind